T0272650

THE AGE OF SCIENTIFIC WELLNESS

The Age *of* Scientific Wellness

Why the Future of Medicine Is Personalized, Predictive, Data-Rich, and in Your Hands

Leroy Hood *and*
Nathan Price

THE BELKNAP PRESS OF
HARVARD UNIVERSITY PRESS
Cambridge, Massachusetts
London, England
2023

Printed in the United States of America

Second printing

This book is meant to educate, but it should not be used as a substitute for personal medical advice. Readers should consult their physicians for specific information concerning their individual medical conditions. The authors have done their best to ensure that the information presented here is accurate up to the time of publication. However, as research and development are ongoing, it is possible that new findings may supersede some of the data presented here.

Many of the designations used by manufacturers and sellers to distinguish their products are claimed as trademarks. Where those designations appear in this book and Harvard University Press was aware of a trademark claim, the designations have been printed in initial capital letters.

Library of Congress Cataloging-in-Publication Data

Names: Hood, Leroy E., author. | Price, Nathan D., author.
Title: The age of scientific wellness : why the future of medicine is personalized, predictive, data-rich, and in your hands / Leroy Hood and Nathan Price.
Description: Cambridge, Massachusetts : The Belknap Press of Harvard University Press, 2023. | Includes bibliographical references and index.
Identifiers: LCCN 2022039356 | ISBN 9780674245945 (cloth)
Subjects: LCSH: Medicine, Preventive. | Health status indicators. | Diagnostic services. | Preventive health services—Technological innovations.
Classification: LCC RA425 .H588 2023 | DDC 614.4/2—dc23/eng/20221007
LC record available at https://lccn.loc.gov/2022039356

This book advocates scientific wellness via healthcare that is predictive, preventive, personalized, and participatory. We dedicate this book to all those who will help solve the very challenging fourth P. Achieving this requires so much more than our efforts alone. We must persuade partners across all aspects of the healthcare ecosystem of the life-changing path before us. We wish you success in optimizing your own journey and hope you will join us in ushering in the age of scientific wellness.

Contents

THE AGE OF SCIENTIFIC WELLNESS

Introduction

This is a book about the future. It's about the promises, pitfalls, and challenges we will have to overcome if we want to take the next big leap in human health. Not everyday progress that comes incrementally, but *exponential* improvement in the human condition. Our deeply held belief, which has emerged from our personal and scientific experiences, is that we are in the first stage of the largest paradigm shift in healthcare since the beginning of modern medicine—a time when the fundamental ways in which we approach health will change so profoundly that we will struggle to understand why we ever did things any other way.

Soon, we will be able to track and optimize the health trajectory of every individual throughout their life. We will be able to keep their bodies healthy and their minds young much longer than we do now. New technologies will make it possible to vastly improve the well-being of every person on this planet. (Yes, *every* person, for we should consider wellness to be a fundamental human right and strive to share these benefits broadly.)

This admittedly bold vision begins with a dismantling of our contemporary medical paradigm, a model that is really "sickcare" rather than "healthcare." In today's world, medical interventions happen long after a disease has taken hold, following a centuries-old medical playbook that goes like this:

1. Wait for something to go wrong.
2. Try to identify what caused the problem.

3. Try to fix it.
4. If it works, try the same approach on the next person.
5. If it doesn't work, treat the complications—or write a death certificate and move along.

Even when the current model for care does "work," significant damage has been done. We have long known that people who contract one disease are more likely to become sick with another. Diabetes increases the risk for dementia and coronary artery disease.[1] Cancer increases the risk of pulmonary thromboembolism.[2] And so it goes. It is true that risk factors for any one disease are potential starting places for many others. But there is another reality that physicians and researchers are beginning to recognize: our body is a "system of systems," in which the perturbations caused by disease in one place can lead to seemingly unrelated problems in other places. It does us little good to view the human body in binary terms as either "well" (in which case medical care is unnecessary) or "sick" (in which case it is). Yet this is how our healthcare systems are built—not just in the United States but around the world. But there is a whole spectrum to wellness, and a long trajectory in the progression of disease to end-stage illness.

This book offers an alternative, a vision of twenty-first-century medicine that might meet and prevail over the vast medical challenges of our time. This vision begins with the dismantling of our current binary framing of wellness and disease in favor of the more nuanced, more accurate, and more commonsensical notion that each person's state of health exists on a spectrum across wellness, transition, and disease.

It should go without saying that the healthcare of the future will seek to optimize the amount of time that we are well (Figure I.1). Yet contemporary medicine does very little in this regard. Most of us, as a result, are less healthy than we think. It is likely, in fact, that the physical and mental youthfulness, vigor, and resiliency that represent the wellness phase often only exists through the first ten to fifteen years of most people's adult lives (perhaps as little as 20 percent of their adult life span).

If you are older than your twenties, it is very likely that you have already slipped out of a state of wellness and are in a phase of transition to disease and aging-associated declines. This phase can occupy a large portion of our lives, and yet it is all but completely ignored in the cur-

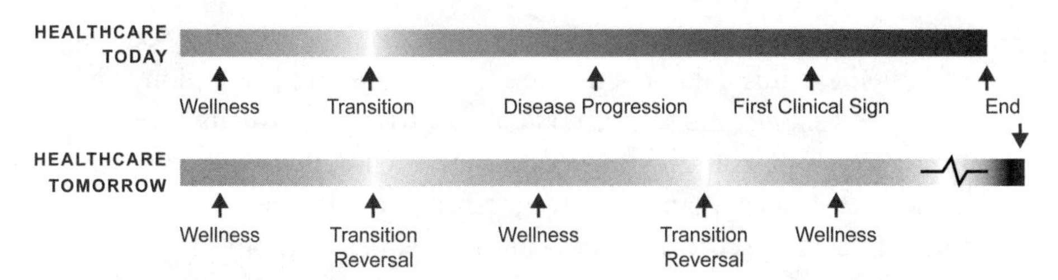

Figure I.1. The healthcare of the future will focus on extending wellness.

rent healthcare paradigm. In fairness, this is partially because it has long been almost impossible to see these early wellness-to-disease transitions clearly. Soon, however, we will have the capacity to identify these transitions and safely intervene, reversing the movement toward disease and returning individuals to outright wellness. We will be able to do this again and again, throughout the course of a person's life.

We (Lee and Nathan) are both optimists, but we do not foresee a future in which we beat back all diseases, altogether and forever. If and when a transition to fatal disease does happen, it should come swiftly—and as painlessly as possible—capping a full, productive, healthy, and happy life of overwhelming wellness. So how do we get there?

Wellness, Transition, and Disease

Our current health trajectories feature an early period of wellness followed by mostly imperceptible transitions, leading to the onset of diseases in the middle of life. (More than one in four adults in the United States has at least one *diagnosable* chronic condition by the age of 44, and about two-thirds will be diagnosed with a chronic disease by the time they reach the age of 65.)[3] So it is that many of us spend half or more of our lives in a state of disease, whether we know it or not.

The idea that drives our new paradigm is to optimize each phase of the human health trajectory—lengthening periods of wellness and improving its quality, detecting and reversing transitions at the earliest possible moments, and greatly delaying (if not altogether avoiding) disease. We will do this through a data-driven process we call *scientific wellness*, an approach that will extend the wellness phase from the twenties to the eighties, and eventually even beyond.

There are three categories of information we need for every person to implement this data-driven vision of wellness and prevention. The first is the *genome*, the source code of life, which is virtually invariant throughout a person's lifetime (save for the case of the mutations in cancer cells). Despite major advances in genomic analysis that have brought down the cost and time it takes to sequence an individual's genome, fewer than 0.01 percent of the people on this planet have had their entire genome sequenced. Even though this only has to be done once, it's a first hurdle on the path to data-driven health. Thankfully, this is now simple to do through numerous affordable commercial offerings, and hopefully soon will be a routine part of healthcare.

The second is the *phenome*, an assessment of your body's status at any point in time during your life as a result of the interaction of your genome, your lifestyle, and your environment. Your phenome changes continuously, and it can be sampled at any time through certain measurements such as the gut microbiome and blood analytes, which are the proteins, metabolites, and other molecules that move through your body and mediate many of life's functions, including energy production, nutrition, and cognition. Because the phenome is constantly changing, it needs to be sampled far more often than the genome—ideally, several times a year—and this is a rather big second hurdle.

The third category is *digital measurements of health*. This might be the easiest of the three, as there are already hundreds of millions of people worldwide collecting this kind of data on themselves, most often through the use of smartphones, watches, smart rings, and other wearables that can track their heart rate, body temperature, respiration, activity level, calories consumed and burned, sleep, menstrual cycles, blood sugar, hormone balances, and more.

With these three categories of data in hand, we can begin to assess the optimal physiology of one's body and brain and detect early transitory phases many years—even decades—before a disease becomes clinically apparent.[4] With this sort of lead time and a positive focus on optimizing health and resilience, we can use these same data to design and target personalized therapies that will end the transition long before disease materializes, when the pathological changes at hand are less complex and more reversible through interventions that are simpler, safer, and less intrusive.

You may have heard aspects of this bandied about in the past by medical practitioners who have suggested that "early detection" is a panacea for all that ails us. But as most physicians will attest, the pitch has always been better than the product, partly because such strategies are aimed at catching symptoms of disease, not signals of the early wellness-to-disease transitions. We have been trying to stop wildfires by watching for smoke on the horizon. But where there is smoke, there is *already* fire. If we get this right—and we cannot overstress how strongly we believe that we can—we will be able to intervene long before the blaze begins. One's health trajectory will become a series of wellness phases punctuated by short transitions, immediately reversed, that will eventually be imperceptible from their general state of robustly good health. For the vast majority of our lives, we will be *well beings*.

Scientific wellness will ultimately allow us to conquer heart disease, the number-one killer in America today. It will make it possible to eliminate or vastly reduce incidents of diabetes or rheumatoid arthritis. It will be our ally in the so-called war on cancer. It will make the scourge of Alzheimer's disease a distant memory. This may seem far-fetched, but we are on the cusp of a time when we will have the capacity to begin to eliminate most chronic diseases—though doing so will depend in part on people making choices in their own lives that will keep them well.

Let us be candid about how we see this revolution impacting future pandemic threats. Infectious diseases are almost always more dangerous—and in many cases exponentially deadlier—for individuals with pre-existing chronic conditions. This was the case with COVID-19, which sometimes killed indiscriminately but far more often took the lives of those who already had one or more chronic diseases, whether or not they had been formally diagnosed. So, scientific wellness will also be key to fighting the infectious diseases that will be an inevitable part of our future.

What would rapid and effective early diagnosis have meant for the deadly coronavirus pandemic? It might have saved hundreds of thousands of lives. COVID-19 was most lethal to those who already had one or more pre-existing conditions such as cardiovascular disease, chronic kidney disease, chronic lung diseases, diabetes, hypertension, obesity, and autoimmune diseases like rheumatoid arthritis and lupus.[5] In the United States, older Americans represented about a third of the early cases, but they accounted for about half of intensive care admissions and more than three-quarters of deaths.[6] Pre-existing conditions exacerbated the death toll. As the coronavirus pandemic overwhelmed healthcare systems around the world, patients who might have been saved with relatively basic critical care—the nearly 100-year-old technology known as a ventilator—were sometimes left to fend for themselves. There are no words to adequately describe the depth of this global tragedy. Yet it could have been ameliorated by earlier diagnoses and a better understanding of the possible disease trajectories.

If some of those pre-existing conditions had been addressed in the earliest stages of the wellness-to-disease process, how many lives could have been saved? Among those who eventually recovered, the path to health could have been so much smoother, and the draconian measures taken by much of the world largely avoided.

When it comes to infectious diseases, the best public health strategy is the one that will finally make a difference in the fight against chronic disease. This strategy is not just to be proactive by a matter of weeks or months; it is to be proactive by years and decades. A healthier population without pre-existing conditions is a population that will be less susceptible to the next pandemic, and the one after that, and the one after that. We, of course, collectively know that health is important, but until COVID-19, few viewed healthy sleep, exercise, and diet as active and pressing strategies to survive or minimize disease. One has only to look at the drastically slower progression of the SARS-CoV-2 virus in Japan, where diabetes is almost nonexistent, and hypertension and obesity are so much lower than in the United States. This is just the beginning of what scientific wellness informed by longitudinal data clouds promises—to detect the earliest wellness-to-disease transitions and, hopefully, reverse them at their earliest detectable stage. Wellness-centered healthcare throughout life is key to achieving these goals.

These bold proclamations come with some caveats. This future will not be realized if we follow a centuries-old strategy of seeking cures that are good for only some people, some of the time. Few people realize it, but the ten most popular drugs in the United States today—from esomeprazole and rosuvastatin to fluticasone and pegfilgrastim—work, collectively, for only about 10 percent of treated patients.[7] Too many people are being subjected to their known side effects without benefit to their underlying condition. We will also fail to realize our goals if we ignore brain health, as is largely the case in contemporary medicine. A larger and larger share of the population is reaching the ninth and tenth decades of life, but often in such a state of mental frailty that these extra years are more of a burden than a blessing. This future will also fail to come about if we continue to treat sickness with the blunt force that even "noninvasive" treatments and therapies wield against disease. The side effects of medications and consequences of many procedures are often described by patients as a "cure worse than the disease." And we will not realize this future if we stubbornly cling to the timeworn idea that a person's chronological age is a suitable stand-in for assessing their state of biological aging—a hard-to-shake notion, even though it is so patently clear that not everyone ages at the same rate.

Once we dismantle these four damaging myths about human health, we can embrace a new standard for medical care that uses each person's genetic profile and phenomic measurements to generate a unique list of "actionable possibilities." In most cases, these proactive behaviors, verified by clinical studies, will either optimize wellness or prevent or forestall wellness-to-disease transitions in the body and brain. When a transition does occur, global, holistic, and data-driven approaches to this disease would inform a precise medical response, using massive data analysis to provide fundamental insights into how to approach therapy effectively for each individual.

All of this will offer the enticing opportunity for each of us to feel confident that our health spans—the years spent living in wellness without disease—will better align to our life spans. Ultimately, a life span and a health span should be virtually the same, meaning individuals would be able to live into their nineties or longer, maintaining an effective state of mental and physical wellness throughout these years.

This may sound like science fiction, but it is not a vision for some distant future. While it will obviously take time to change mainstream medicine, and every one of us will have to participate in our own health journey, we fully believe that the primary actions that must be taken to realize this goal can be completed in the next fifteen to twenty years. Indeed, many doctors and scientists are already taking meaningful steps toward applying the principles of scientific wellness to patient care and biomedical research, including the recent rapid growth of clinics dedicated to personalized medicine, functional medicine, integrative medicine, or healthy aging. They are embracing data from the major determinants of each person's health—their genome, their lifestyle, and their exposures to the environment—to build a kind of care that is predictive, preventive, personalized, and participatory, what Lee was the first to call "P4 medicine." And they are finding solutions for the major challenges of contemporary medicine: poor quality, exploding costs, the rapid aging of the patient population, and the dramatic increases in the number of individuals with one or more chronic diseases.

We will dive into some of these challenges throughout this book. Our goal is to help you see the power of scientific wellness, understand what it will take to build on nascent successes, and optimize your own individual health to ensure a long, productive, and healthy life. In fact, we suspect that what you learn in these pages will fundamentally change your view of what is possible for your own health. Our hope is that it will also help you see this way of thinking about health not simply as something that you can benefit from but as a structure for healthcare that can benefit everyone.

Getting to that point requires that we recall that medicine is about people. It's about patients and physicians. It's about researchers, healthcare administrators, and coverage providers. The current healthcare paradigm treats these as competing interest groups, but scientific wellness offers us an opportunity to focus everyone's interests toward a new and mutually beneficial goal—a new standard of personalized medicine built for the twenty-first century and beyond.

We will not tell you that we can immediately shake the foundations of global healthcare and get everything right. We will not pretend that disruptive innovation is not going to be disruptive. That would be foolish. Where we see challenges ahead, we will name them and offer possible solutions. When we do not yet know the solutions, we will say so. It will be difficult. If it weren't, it wouldn't be revolutionary.

Both of us are actively participating in processes that will bring this vision to pass, and we both believe we'll be around when it comes. For Lee, who is now in his mid-eighties, it might seem like a wistful fantasy. But we can see it coming, faster and faster, just over the horizon. To understand our confidence, perhaps it would be helpful to say a bit about the past.

We wrote this book together, and before we turn to the medical breakthroughs of the last ten years and what is coming next, we will share a bit about how we got here. For Lee, it all started a long time ago, in the small town of Shelby, Montana, thirty miles south of the Canadian border.

Lee's Story: A Life in Science

There is a preconception that small, rural towns are poor places for children to grow up if they want to formulate big, world-changing ideas. The information revolution—which made it possible for anyone, anywhere, to access endless information—should have put this idea to rest a long time ago, but from my experience, it was never true to begin with.

While I think a lot about the future, I am not immune to the nostalgia that is common to people of my age. My thoughts often turn to my grandfather's ranch in the shade of the Beartooth Mountains and to the town of Shelby, where I attended high school. When he wasn't on his ranch, my grandfather built and managed a geology camp in the Beartooth foothills, where Ivy League professors brought their students for summer courses and projects. I learned a lot about science from the students and faculty at that camp. In my junior year of high school, I completed a geological map of an oil-producing anticline in northern Wyoming that got me invited to the Westinghouse Science Talent Search in Washington, DC. It was the first time I'd left Montana, and I went by myself on the Great Northern Railway. I was awed by the brilliant students I met in the capital and returned to Montana determined to join their ranks.

My father worked for Mountain States Bell, managing the construction of a series of communication microwave repeater stations across the state. He was a superb engineer, and he taught weeklong summer courses to his employees in general aspects of electrical engineering. He encouraged me to take these courses—mostly, I think, so he could show me off to his employees. I participated with reluctance, as I would rather have been out hiking and mountaineering. In retrospect, these courses changed my life. They taught me to think about biology in terms of engineering systems and circuits—a concept that was later very useful for my commitment to developing new technologies for biology and the new discipline of systems biology.

Shelby High School had only 146 students, but it had just about everything any other school might offer. I played oboe in the school band and was co-editor of the yearbook. I acted in plays and traveled the state as part of the debate team. I served in student government and played quarterback on a football team that was undefeated for two and a half years. And I was blessed with some of the best teachers I would ever have. They treated me as a colleague and broadened my intellectual horizons. They challenged me to think about science as a career with a sophistication I never would have achieved alone.

My most formative experience was helping my chemistry teacher, Clifford Olsen, teach a sophomore biology class. I taught directly from articles out of *Scientific American*, and in the spring of 1956, I taught a lesson I developed from an article on the structure of DNA—only three years after it was discovered by James Watson, Francis Crick, Maurice Wilkins, and Rosalind Franklin. I didn't understand much about it, but the idea that the core of biology was centered around this beautiful molecule fascinated me then as it does to this day.

DNA intrigued me for another reason. My brother Glenn, who was six years younger than me, was born with Down syndrome, and when I learned about DNA, I wondered if it might play a role in his condition. My father and mother were split on how best to care for him. There was a state home for children with Down syndrome in Boulder, a good four hours away, in central Montana. That's where Dad thought Glenn would get the best care. Mom wanted her boy to be at home. Dad, as usual, won the argument, and it was probably the right decision; Glenn flourished, and in his teen years moved to Hardin, Montana, where he spent the rest of his life. Glenn went on to own his own home and hold three jobs simultaneously for much of his adult life.

I respected my brother and desperately wanted to understand Down syndrome. I remember asking our physician and my parents about the cause of Glenn's condition. No one had an answer. And I was always fascinated by questions without answers.

The genetic cause of Down syndrome, a duplication of all or a portion of chromosome 21, was finally discovered in 1959. By that time, I was at the California Institute of Technology (Caltech), following my mentor, Mr. Olsen, who had attended university there during World War II as a Navy meteorologist. He was so impressed with the school he had vowed to send any good science student he had there. At Caltech, I was surrounded by outstanding students, many of whom were already quite a bit further along in their math and science than I was. I had to work hard to catch up.

Caltech gave me a superb technical education in math, chemistry, and physics. My professors included Linus Pauling, the only man ever to win two unshared Nobel Prizes, and Richard Feynman, another Nobel Prize winner, whose work was foundational to quantum mechanics. Both were excellent and inspiring teachers. My biology training was exceptional, too, but it focused almost entirely on microbes, plants, and viruses. I was passionately interested in human biology and disease, and decided to go to medical school in the hope of later doing research in human biology. At Johns Hopkins Medical School in Baltimore, I became transfixed with molecular immunology and the question of how our immune systems could protect us against so many different types of pathogens. One approach to this problem was to study the structure of antibody molecules, a major component of the human immune response. I decided to pursue my PhD in this field, returning to Caltech so that I could work under William (Bill) Dreyer, who had made fascinating discoveries in this area of science.

Dreyer was an unusual biologist because he had a deep interest in technology. He gave me two dicta that have guided me ever since: (1) if you want to practice biology, practice it at the leading edge, and (2) if you want to transform a field, invent a new technology to interrogate it.

The Vietnam War dictated the next phase of my life. As an MD, I had two choices: I could join the military or serve in the public health service. I chose the latter and was assigned to the National Institutes of Health (NIH), where I met many young scientists who would become leaders in the field of medicine over the next fifty years. I also learned to manage my own laboratory. But mostly I decided what I wanted to do in the next stage of my career: study human biology and disease.

And so it was that I headed back across the country to Caltech where, in 1970, I became an assistant professor in biology. I had decided to focus on two areas of research. The first was molecular immunology, a leading-edge topic area in biology at that time. It was the complexity of this field, which deals with disease responses mediated at the DNA level, that so intrigued me—all those challenging questions without answers embedded in the complexity of human biology. But I immediately saw a problem: we didn't have the tools to approach many of these questions. In my emerging view, biology was an information science; without ways of measuring this information, we were lost. This led me to my second focus, the development of new technologies to assess the four major types of biological information: DNA, RNA, proteins, and biological networks (per Bill Dreyer's second dictum, above, which identifies technological innovation as the path to transformative research).

Medicine in those days, and even today, was very much like the parable of the elephant and the six blind men. In this ancient Buddhist story, each man feels a different part of the elephant and comes to a completely different conclusion about what the animal is based on what he has felt. The man who grabs the trunk believes he is touching a snake, while the one who grabs a leg thinks he is touching a tree trunk. Limited by the symptoms of illness, physicians were like the blind men. I came to several conclusions in my time at Caltech. First, I felt it was important to generate lots of data on each individual, for buried in the data were the keys to deciphering human complexity. Second, I became convinced that blood was a window into viewing wellness and disease, because it bathes all organs and receives protein signals from each organ that can reflect the internal health state of that organ. In principle, the wellness-to-disease states of all organs can be assessed from blood. It wasn't until recently that we had the tools to start doing this and understand all that we could learn. Moreover, at the time of my studies, we didn't have the language to convey the complexity of all this biological information or understand how to think about it. We weren't just blind; we were speechless, too. Later, systems biology gave us a language for beginning to decipher human complexity.

Around the time I was coming to these realizations, in the early 1970s, I read *The Structure of Scientific Revolutions* by Thomas Kuhn, which described paradigm changes in physics. These shifts—revolutionary new ways of thinking about or practicing a discipline—are difficult to conceptualize and even harder to achieve. That's because discoveries that come to be recognized as paradigm-shifting moments almost always face staunch resistance initially. Scientists, like the rest of us, are generally reluctant to give up long-held beliefs and accept new ideas.

In spite of the resistance—or perhaps because of it—I have been fortunate to have been involved in a number of paradigm-shifting moments in my career: changes in biology, medicine, and technology that we now see as revolutionary.[8] In homage to my boyhood home, I've come to think of these as "Big Sky moments." They led me to a new approach to biology founded on the realization that humans employ complex biological systems to carry out the normal functions of the body, and that focusing on any one of them to the exclusion of the others will not get us very far.

Bringing Engineering to Biology

Twenty years after the discovery that DNA molecules exist in the form of a three-dimensional double-stranded helix—a finding that gave us the conceptual insights we needed to understand what our genome does—we still didn't have effective tools to explore this code. DNA is a digital code with a four-letter alphabet—the bases G, C, T, and A. By the time I started teaching at Caltech, we knew that DNA was the source code of all life, and we were beginning to use this nascent realization to better understand how humans develop from a single fertilized egg into an adult human. For this to happen, all 20,000 units of DNA encoded in our genes have to be copied into another four-letter language, messenger RNA (or mRNA). These single-stranded molecules are then translated into the twenty-letter amino-acid language of proteins—the functional machines of life and a key part of the biological networks through which our bodies operate. Since the information they carry is encoded in subunits in DNA, RNA, or proteins, one must be able to determine the order of these nucleic acid bases and characterize the order of amino acids in a protein to understand the nature of this linear information. This is what we now call sequencing.

Pehr Edman, a brilliant Swedish biochemist, found a way to sequence proteins in the 1940s. He was the first person to build an automated sequencer, which sped up the process considerably. His tools and methods, however, required lots of proteins and could not generate long sequence reads. When I returned to Caltech, sequencing remained a long and labor-intensive process. Since my initial training was in protein chemistry, I thought I might be able to develop an instrument that could sequence proteins more efficiently, offering longer sequence reads with far less starting material. If I could achieve this, proteins available in very small quantities could then be characterized, their genes cloned and sequenced, and, hopefully, their missions discovered.

This was not simply a challenge of biology but also one of chemistry and engineering.

There was a lot of pushback to such cross-disciplinary work in those days. For some biologists, it was anathema. In 1973, Caltech's chairman of biology, Robert Sinsheimer, came into my office and asked me to give up on technology development. "Your field is molecular immunology," he reminded me. "That's where your focus should be."

I told him I wouldn't be altering my goals and waited to see what the fallout might be. Sinsheimer later told me he was delivering the message on behalf of the school's senior biology professors, who felt it was inappropriate for me to practice engineering in a biology department. If that was my focus, they had suggested, I should be moved to the engineering department. To his credit, Sinsheimer never sought to have me transferred.

The resistance didn't just come from Caltech. When I sought support for my team's automated DNA sequencer from the National Institutes of Health, my first two grant applications received priority scores that were among the worst I would ever get. The reviewers offered comments like "this approach is impossible" or "graduate students can easily do all of the sequencing needed." They seemed to have little appreciation for the exponentially increasing amount of sequencing that would soon be the bedrock of so much biological and medical research—and even less for how talented graduate students could most effectively use their time.

I wish I could say only a few outliers held this view. I remember talking with Jim Watson, the co-discoverer of the structure of DNA, in the mid-1980s about automated DNA sequencing. "Why are you putting so much time and resources into this project?" he asked me. I tried my best to explain, convinced that automated DNA sequencing would transform biology, but I don't think I succeeded. "Just remember," he said sarcastically, "that there are a billion people in China, and if each one of them sequenced just three bases of the human genome, it would be finished."

My initial efforts were only marginally successful, and there were times when I thought the naysayers might be right. But in the mid-1970s, a brilliant chemist and engineer, Michael Hunkapiller, joined my lab. We collaborated with Bill Dreyer, my former PhD mentor, to develop an idea that Dreyer had conceived for protein sequencing—a liquid-gas phase instrument. This approach eventually led to a technology that could create long sequence reads with 200 times less protein than previously required.

Over the next twenty-five years, my collaborators and I worked across disciplinary boundaries to develop six different instruments to analyze and synthesize DNA and proteins in various ways. These included automated DNA and protein sequencers, automated DNA and peptide synthesizers, the very large-scale, inkjet-based DNA synthesis technology, and the single RNA molecule NanoString analysis technology.[9] The automated DNA sequencing instrument approach employed four different fluorescent dyes, one for each DNA letter. With a synthesis sequencing technique that allowed us to visualize each base in a DNA sequence, we were able to convert the order of colors obtained by laser scanning into the sequence of the DNA fragment by a computational transformation. The four-color DNA sequencing chemistry we developed at Caltech has been the cornerstone of automated DNA sequencing for the past thirty-five years.

At last, we could read this sacred text, the code of life—and we had the combined forces of molecular biology, chemistry, engineering, and computer science to thank.

The Human Genome Project

Now that we had the tools to sequence DNA more efficiently, we could focus on a previously impossible goal: determining the order of the four nucleotide bases of a DNA strand in each of the twenty-three pairs of human chromosomes—what came to be known as the sequence of the genome. This was a daunting task, as about three billion DNA letters make up the human genome, and chromosomes range in size from about 50 million to 175 million base pairs.

This was not a mountain to climb just for the sake of being the first. For decades, scientists had been dreaming of what they might learn if DNA were ever to become more easily readable. Once the order of bases in the human genome was determined, many believed, we would have the information we needed to begin to understand how different genes are expressed in different tissues—to understand why a liver cell is different from a brain cell and so forth. There was also a widespread hope that one could correlate defective genes with different disease states and better understand the mechanisms of disease so that we could develop targeted therapies to deal with them more effectively. Writing in the journal *Science* in 1986, Renato Dulbecco of Caltech suggested that sequencing the human genome might be the key to understanding cancer.[10]

All of this was easier to dream about than to accomplish. While I was initially skeptical, I was intrigued enough to attend the first meeting on the Human Genome Project in the spring of 1985 on California's Central Coast, where Robert Sinsheimer had become chancellor at the University of California Santa Cruz. Sinsheimer had invited twelve experts in genetics, human biology, and technology to assess the merits of an effort to sequence the human genome in its entirety for the first time.

Bart Barrell and John Sulston, both biologists, had flown from Cambridge, England, and visionary biotechnologist Hans Lehrach had traveled from Heidelberg, Germany. The Massachusetts delegation included the biologist David Botstein, biochemist and physicist Walter Gilbert, and Leonard Lerman, a geneticist. Helen Donis-Keller, whose company, Collaborative Research, was working to produce one of the first linkage maps of the human genome, was also in attendance. David Schwartz, a chemist and geneticist, had arrived from Columbia University. The shortest trips were made by the California contingent, which included George Church, a now-famous geneticist who was then at the University of California San Francisco; Ronald Davis, a biochemist from Stanford; Michael Waterman, a biologist and mathematician from the University of Southern California, and me.

Together, we debated the merits of attempting such a feat. Even with an early-stage automated DNA sequencer, we agreed it would be extremely challenging—and tremendously expensive. We were split, six to six, on whether it was a good idea. One major objection was that the project represented "big science," and there was a fear—not unreasonable—that it could steal resources away from the more targeted biological research that was then the cornerstone of biology. No final decision was made in those three days, but the excitement was palpable, and those of us who were most in favor of the project began crafting plans to build support.

As I went out into the scientific community to talk about the project, I found that some 80 percent of the biologists I spoke to were opposed to the idea. So were leaders from the National Institutes of Health. That's right: the NIH, the largest funder of biological research in the United States, wasn't supportive of funding the Human Genome Project at first. The sources of resistance were broad in scope: technological, social, ethical, legal, organizational, economic, and political. I remember giving a Friday night lecture at the Marine Biological Laboratory at Woods Hole, Massachusetts, in 1986. These lectures, which date back to the 1890s, are famous as sounding boards for new ideas in science. The hall was filled, and I was initially satisfied that I had made a sound case for the Human Genome Project, and for using the power of automation to make it happen.

This impression did not last long.

"You say 'automate this' and 'automate that,'" a prominent colleague offered at the onset of the time reserved for questions. "I ask you: Where is the humanity in your science?"

I was taken aback. I had thought it was obvious that automation would bring humanity back into science by freeing researchers from needlessly mundane tasks. Maybe, I thought, this critic was just an outlier—but the reactions actually went downhill from there. I had arrived in Woods Hole just before the lecture. My host, caught up in the hostility, went off with others after the lecture and left me alone with my thoughts and my luggage. This was in the days before Google could point you in the right direction, and I had to ask a janitor where my hotel might be.

Over the next few years, as funding opened up and institutional support solidified, most of us working on the Human Genome Project heard from many scientists who expressed concerns similar to those I'd heard in Santa Cruz, worrying that this enormous project would take resources away from smaller research efforts. And to be fair, there was no lack of important "small science" research being done at that time, with a senior scientist and perhaps a handful of colleagues focused on a single gene or protein. I wanted them to think about what they could accomplish with an entire genome, but instead they were focused on what they could accomplish with the sort of money we would need to complete that genome sequence.

The project might never have gotten past this resistance had it not been for a seemingly unlikely source of support: the US Department of Energy (DOE). In the wake of the development of the atomic bomb, the DOE had directed some of its research efforts to understanding the ways in which radiation causes genetic damage. Without an entire human genome, argued visionary Charles DeLisi, who was then the director of the department's Health and Environmental Research Programs, it would be impossible to answer such questions. The DOE's initial funding supported the development of key technologies needed to begin the Human Genome Project.

I had thought that powerful ideas would open minds to the opportunities inherent in the project. In fact, it was the tools that swayed most scientists. As the physicist Freeman Dyson would later write in *Imagined Worlds:* "New directions in science are launched by new tools much more often than by new concepts. The effect of a concept-driven revolution is to explain old things in new ways. The effect of a tool-driven revolution is to discover new things that have to be explained."[11]

Still, the big-science-versus-small-science debate was a tremendous hurdle. It was US Senator Pete Domenici and his colleagues who helped us see this as a political challenge and conquer it as such. The Republican from New Mexico initially supported the genome project because it would benefit the two large national laboratories in his state, Los Alamos and Sandia. Later, he advocated for the project based on the idea that it would give the entire nation a competitive edge in biotechnology. The key to gaining broad support, Domenici recognized, was to go after *new* money, so that small-science advocates couldn't argue that resources were being taken from them.

By the late 1980s, there had been a dramatic shift in excitement for the genome project. Perhaps the most determinative issue was the fact that a National Academy of Sciences committee (initially including advocates and opponents) wrote a report unanimously backing it. NIH soon jumped in with enthusiasm and money. That was important, because our initial prediction that the project would be expensive was correct. All told, it cost about $3 billion to produce the first human genome sequence.

In 2013, the Battelle Memorial Institute estimated that our initial investment of $3 billion had delivered more than $800 billion to the US economy—a return on investment previously unparalleled in biology.[12] Some argued at the time that this was a poor way to gauge success, saying that the true return on investment could only be measured in health outcomes, but either way, in my view, this was money well spent. Determining the sequence of the human genome gave biologists access to human genetic variability. And that, in turn, offered us the ability to do something that could never have been done before: we could begin to correlate the variations in our genes with health and disease outcomes. Human genome sequencing was originally "overhyped," leading to disappointment in the speed of its translation to medicine. Still, the project transformed many different fields in biology, and I believe the most important human-health returns are only just now beginning to be realized.[13]

Cross-Disciplinary Biology

If the idea of biologists and engineers joining forces created a storm of controversy, what came next was a veritable typhoon. As my lab evolved to handle the complexity of developing multiple types of instruments, it became increasingly necessary to interface with an even broader set of scientists with diverse skills. To create leading-edge biological technology, you had to gather under one roof all the different specializations required for a wide range of highly complex technical projects beyond biologists: chemists, computer scientists, engineers, mathematicians, physicists, and physicians. You had to put them in proximity to one another and create the conditions for serendipitous interaction. You had to teach them how to understand one another's languages. And you had to treat them the way a good football coach treats a team—respectful of each person's contribution to the whole and mindful that everyone is an expert on their own position. It is teams like this that are now driving the development of groundbreaking technologies and the creation of multilayered visualization, measurement, and computational tools capable of analyzing big data. These teams are poised to revolutionize medicine by generating tools that are capable of "seeing" new dimensions of biological data, enabling biologists to formulate new hypotheses about human health and disease.

The development of sequencing tools would not have been possible without teams like this. Across the scientific world, agile minds were thinking alike. In 1987, the National Science Foundation launched its Science and Technology Centers program, an initiative aimed at tackling big science problems through the recruitment and integration of scientists and engineers. I applied for a grant to create one of its first programs; we called it molecular biotechnology. It was the most effective grant I ever received from the federal government, in that much of the money was discretionary and available to catalyze immediately new opportunities.

Microsoft founder Bill Gates, who was then and is now a tremendously generous supporter of innovations that can improve human lives, was an early believer in cross-disciplinary work like this. He played a vital role in helping me realize my vision for building such teams. Disheartened by the frequent lack of support from senior colleagues at Caltech, I had been contemplating whether my work might be a better fit in another place. In 1992, Gates made it possible for me to start the first academic research unit in the nation dedicated to cross-disciplinary health science research and development at the University of Washington (UW) Medical School. The new department came to be named Molecular Biotechnology.

The provost at Caltech was bewildered when he learned of my intention to move. "Lee," he told me, "you should go see a psychiatrist. No rational person would leave Caltech for the UW." That sort of academic arrogance was off-putting. I didn't care about status; I wanted to work in an environment where scientists were enthusiastic about what we could be doing *together.*

If you look at a health science journal from decades ago, you might notice that most articles had one, two, perhaps a few authors. That's not just because senior investigators used to be quite a bit more selective when it came to sharing credit. It's also because, in the days of less complex tools and fewer measurements, the breadth of knowledge required to conduct experiments was smaller. Not so long ago, a single person could advance an entire field based on what they saw under a microscope in a petri dish. For many areas of biology, those days were long gone by the time the first human genome was sequenced. Although there is always room for brilliant, single, individual ideas.

These days, health science articles authored by only one or two people are rare. It takes teams working across disciplines to integrate and analyze diverse types of information to do science that moves the needle. Today, advances in medicine often take *systems* of people, working together as part of an interconnected network. A new way of thinking about and practicing biology is taking shape.

The Rise of Systems Biology

Despite the tremendous advances that gene and protein sequencing offered us, most of the research world was still locked in classically reductionist research. We studied biological systems one gene at a time, one protein at a time, one disease symptom at a time, ignoring the ever-mounting evidence that there are very few processes in our bodies in which A causes B, and B causes C, and so on in a simple, linear manner. Rather, an increase in A might start a chain reaction in which we see a decrease in B, which prompts a switch between C and D. D, however, is consumed by A, causing A to plummet, but when B rises as a result, C and D cannot switch back, because D has been used up by A. All of this is to say that biology generally operates as complex systems rather than linear pathways.

We may have identified most of the 20,000 or so genes in the human genome, but we are still working hard to understand what they do individually and, more important, how they operate in networks where one component of the network may influence many others.[14] We are only just beginning to assemble genes and their proteins into biological networks for regulating the expression of genes, metabolism, and the functions of protein networks. We are barely able to understand the dynamics of these systems, but that doesn't change the fact that, without a systems approach to biology, we are helpless when it comes to truly understanding any biological process.

A useful analogy for understanding systems biology, one that we have borrowed from Yuri Lazebnik, is to think about how one might go about figuring out how a radio works.[15] One could break down the radio into its component parts and study what each one does individually. This is essentially what biologists were doing before the advent of systems biology—studying the individual components of life: genes, proteins, or metabolites (the key components of our metabolism) one or a few at a time. That's a good start, but the parts of a radio don't work independently from one another. Most of them don't do much unless they are connected to other parts.

A systems approach to understanding the radio would require: (1) defining the parts, (2) connecting the components to their wiring circuits, (3) carrying out experiments to understand what the individual circuits in the radio do, and (4) understanding how all these circuits come together to create the different functions of the radio. In a similar manner, scientists engaged in systems biology conduct experiments to determine how each part of the human body is connected to other parts and to the whole. This includes, for instance, investigating how genes and the proteins they encode are joined into biological networks, considering what the functions of the individual networks are, and examining how the various networks are interconnected. We do this through experiments that determine the operational dynamics of the systems in an effort to understand how they give rise to disease.

I wanted to build a systems biology center on top of a cross-disciplinary framework at the University of Washington. Alas, the bureaucracy of a large state university made it a challenging task. When I asked for space for the growing computer needs of my department, the head of university computing informed me that a janitor's closet was all the room a biology department would ever need for computing. He failed to understand that biology was an informational science with exponentially increasing computational needs. Later, I persuaded a prominent cell-surface chemist from Pennsylvania State University to join my faculty, but the dean informed me that cell-surface chemistry was an inappropriate subject matter for a medical school faculty member. When I sought to use the Gates endowment to bring this brilliant scientist to my department, I was overruled.

It wasn't all frustration. Just as I had been able to blossom as a researcher at Caltech, the University of Washington was a tremendous venue for trying out new ideas. Ultimately, though, if you are someone who is used to pushing the envelope, you're liable to push it further than the bureaucracy of a large state university will permit, so I eventually concluded that if I wanted to create a systems biology institute, I would have to start my own.

In 2000, I left the University of Washington to start the Institute for Systems Biology (ISB), an independent, nonprofit research institute that was the first of its kind. Systems biology represented a holistic and integrated approach to biology and medicine.[16] At ISB, we built a framework for the cross-disciplinary approach we created at the University of Washington, adding to it a dynamic vision of biology as an intricate web of processes that impacts and energizes whole biological systems.[17] Our work challenged some of the very foundations of biological research, raising concerns about what happens when you take cells out of the body to examine them in test tubes or petri dishes. We found that removing cells from their natural environment often changes their behaviors in fundamental ways. You can imagine how well this went over with researchers who had spent their entire careers working on studies that began by looking at a part of something instead of the whole.

I remember talking with one of my friends at Massachusetts Institute of Technology (MIT) around the time we started ISB. He was brutally frank in his assessment of this new systems approach. "It is all hype," he told me. "It will come to nothing." I suspect by "hype" he meant that large global approaches were approximate, error-prone and unlikely to yield any fundamental insights. This has not turned out to be the case. Indeed, today there are more than 100 systems biology institutes, departments, and centers across the world, and these institutions are tackling biological and medical questions that previously couldn't even have been asked.

Among our early successes, we learned how galactose metabolism, which delivers energy to the body and is key to early human development, functions in yeast, and we used this process as a demonstration for the fundamental approach of systems biology.[18] Our work on galactose enabled us to determine how the different elements (nodes) in the metabolic network influenced one another, giving us an integrated view of how humans metabolize and use galactose (a form of lactose) to store energy as adenosine triphosphate, or ATP. This advance gave us an early indication of how potentially powerful the systems approach to biology could be, offering the promise that it might help us understand complex aspects of human physiology and raise the possibility that we might even be able to tinker with it to optimize its functioning.

We developed a model, published in *Cell* in 2007, that made it possible to accurately predict the behavior of a simple microorganism, a halobacterium that lives in high saline conditions, given knowledge of its biological networks and outside signals. Our model gave us key insights into the symbiosis of different life-forms and their environments. We even managed to predict the halobacterium's responses to environmental signals that it had never seen before.[19] It was a modest beginning leading to a future where deep insights might be gained from the perturbations of human biological networks.

We learned how the brain's transcriptional networks change during the initiation and progression of a neurodegenerative disease induced by prions in mice, commonly known in humans as "mad cow disease." The most fundamental insight that came from this study, which we published in 2009, was a realization that disease is very simple at its inception and that disease-perturbed biological networks become increasingly complex, mitigating against the likelihood of successful late treatment by any single drug.[20]

Almost ten years after the conversation in which my colleague from MIT tried to warn me away from the "hype" of systems biology, he coauthored a National Academy of Sciences report in which he described the biology of the future, essentially, as systems biology.[21] He is not the only one to have changed his tune. These days, when scientists are asked to speak about the future of biological research, they generally point to a future that is holistic, data-intense, integrative, dynamic, and network-based—all systems approaches. The simple idea driving this progress is that changes in mRNA, proteins, metabolites, lipids, and other small molecules can be traced back to the physiological networks that control them, and that these can give us vital insights into human biology and disease.

A Demand for Personalized Medicine

As we applied systems thinking to the challenge of moving our science from the lab to the clinic, one thing became very clear: you cannot hope to understand a biological system that is chaotically broken though disease if you do not understand its normal operational state. This means you cannot hope to understand a state of disease if you don't understand a state of wellness. Once we started investigating the wellness side of the human health equation, we began to fundamentally alter the way we viewed medicine. Why should medicine be all about fighting disease? Why shouldn't it, first and foremost, be about keeping people well?

We can and should battle diseases when they happen. This fact is not debatable. But increasingly, it is becoming clear that we have the tools and knowledge to optimize wellness and keep disease from happening in the first place, but we just aren't using them effectively. This realization, together with systems thinking, led me to the conceptualization of P4 medicine—healthcare that is predictive, preventive, personalized, and participatory.

The marketing gurus at "Healthcare Inc." love all of these words. Alas, truth is no prerequisite to branding. That's especially the case when it comes to personalized medicine, which has been widely adopted as an advertising slogan but is rarely embraced as an effective practice. Large healthcare systems tell their customers they'll treat them as individuals, but that's not what most patients experience.

But it is, unquestionably, what people want. The past twenty years have seen a notable shift for many patients from large-scale health providers to caregivers who practice what is often called "complementary and alternative medicine." In some cases, these providers offer well-researched practices like correcting measured blood deficiencies in vitamins or nutrients (e.g., iron, omega-3 fatty acids, or vitamin D or B_{12} deficiencies with clinical consequences), but they also can address emerging areas of health like the gut microbiome that simply have not yet been integrated into mainstream care. Some physicians have begun to recommend "alternative" services to patients who have not found healing for their chronic illnesses through conventional medicine.[22] In many situations, however, these alternatives are nothing less than outright quackery, with the best-case scenario being a placebo effect and the more common experience leading to no improvement or a decline in patient health.

When mainstream doctors try to understand what drives so many patients to try alternative medicine, we often tell ourselves that desperate people do desperate things. I believe that this is only a small part of the answer. The bigger force at play is each patient's desire for care that feels it is actually meant for them as an individual. And despite the valid criticisms of some purveyors of alternative medicine, this personalized and wellness-focused experience is what patients feel they are finally receiving when they move away from the dehumanizing experience of robot-facilitated customer support lines, packed waiting rooms, insurance-dictated treatment regimens, and doctors who are in and out of their lives in minutes and prescribing medicine by the playbook while paying little attention to the patient's individual condition.[23] We see the scientific wellness revolution as delivering the kind of care people clearly want predicated by the deep scientific evidence base they deserve.

So why can't conventional care providers walk the talk when it comes to personalized care, winning back these patients who are so anxious to be treated as individuals? Simply put, it is because arriving at care that is truly personalized but also scientifically validated requires a data-driven systems approach to wellness. It takes medicine that is predictive and preventive—which, as we will see, is only possible with a vast array of measurements, assessing many of the hundreds of biological networks functioning within the body—to identify what wellness is uniquely for each person. That baseline is essential if we are to determine what the early stages of a transition to disease might look like in an individual. To get there, we need to know each person's genome, analyze their ever-changing phenome, and take into account relatively continuous measures of digital health. Only then will we have the depths of data needed to predict and prevent disease through interventions targeted at the earliest detectable transitory phase. And that, as you can imagine, isn't going to be cheap. (Not *yet*, anyway—we'll return to this.)

The health maintenance organizations that provide the overwhelming bulk of patient care right now would rather talk about personalized care in their marketing campaigns than do the work that actually makes these words a truthful representation of what is happening in their clinics and hospitals. After all, truly predictive, preventive, personalized, and participatory medicine would incur most of the costs well ahead of the time when the bills are traditionally paid (i.e., after the transition to disease has already occurred). The resistance to this idea feels like an echo of what I heard when I first sought to convince biologists of the importance of the Human Genome Project.

But our vision was realized then, and I believe the same will come about for P4 medicine. Indeed, it is already happening. My colleagues and I have offered personalized care of the kind I am describing for thousands of individuals and, as we shall soon see, it works beautifully. That's one of the key reasons I believe this is a paradigm shift that, while by no means complete, is *already* taking place. Indeed, there are several programs, including NIH's million-person All of Us study, Singapore's million-person project, and the UK's Our Future Health project (which includes five million people) that are beginning the research necessary for data-driven health. Later in this book, we will discuss how you can join this process over the next five to ten years.

It should be pointed out that the first three Ps of P4 medicine—prediction, prevention, and personalization—are driven by science, whereas the fourth P, participatory, is a psychological, sociological, and economic challenge. How do we persuade patients, physicians, healthcare leaders, regulators, and all members of the healthcare ecosystem to participate in a paradigm change that will be the largest ever in medicine? The fourth P is far and away the biggest challenge for twenty-first-century medicine.

Now you now know a great deal of my story. I am a biologist, entrepreneur, and inventor whose automated DNA sequencer was the enabling technology for the Human Genome Project. To some, I am better known as the champion of P4 medicine and founder of proteomics and systems biology. But throughout all of these years and innovations, I was never acting alone. I owe a lot to the fantastic people who have worked in my lab over the years and to many collaborators around the world. In 2005, five years after founding the Institute for Systems Biology, I was approached by a talented young bioengineer, Nathan Price, who was just finishing his doctoral work with Professor Bernhard Palsson at University of California San Diego, the world's leading expert on modeling metabolism at what we call the "genome scale."

Nathan's Story

Nathan Price, who is almost forty years my junior, is an expert on the intersection of big data with biology and medicine. As a graduate student, he built computer models of the vast repertoire of biochemistry in living systems in an effort to answer questions like "How does your body convert the food you eat into energy and all the components needed for your body to function?" These were heady times—the human genome (and other genomes) had been sequenced, and it was possible for the first time to build comprehensive maps of biochemistry, with a catalog of how many enzyme-encoding genes there actually were.

Nathan thrived in Palsson's lab, where he felt he could wake up every morning and ask, "What is the most valuable thing I can do today to advance science?" and dedicate the rest of his day to doing that very thing. Palsson told me Nathan had published the most peer-reviewed papers in the history of the department and, quite unusually, he had already accepted an offer during his final year of graduate school to start his lab in chemical and biomolecular engineering at University of Illinois Urbana-Champaign, one of the top ten departments in the country. But first he wanted to learn more about systems biology and P4 medicine, and that drew him to visit me in Seattle. It wasn't long before he was contributing in important ways to our work at the Institute for Systems Biology. He identified new diagnostic biomarkers for distinguishing cancer types and guiding therapy choice, and invented methods for systems biology data analysis.

After two years at ISB, Nathan started his own lab at University of Illinois Urbana-Champaign. His lab centered on systems biology, and this core expertise gave him a chance to connect with scientists working across a wide range of fields. He led the computational analysis on projects in genetics and social behavior with Gene Robinson, the complex dynamics of biological systems with Nigel Goldenfeld and Carl Woese, metabolic engineering with Huimin Zhao and Hans Blaschek, and the biology of methanogens and their role in climate change with Bill Metcalf and Zan Luthey-Schulten. Although his focus was on personalized medicine and synthetic biology, he also worked with researchers engineering microorganisms for energy sustainability projects.

The University of Illinois is one of the great research universities in our country, with a first-rate engineering school and a storied legacy of thirty Nobel laureates. But Nathan came to believe that if he really wanted to play a role in shaping the future of medicine, he needed to move to the center of the action—and that led him back to ISB.

During his time at Illinois, we stayed in close contact and collaborated on a number of projects in systems medicine. In two years, we developed diagnostic markers for cancers, analyzed disease-perturbed networks, and invented a number of computational tools to make these things possible. We committed to a biweekly call on Tuesdays at 6 a.m., wherever I happened to be in the world, and we spent a week together each summer at my retreat in Montana, where we would talk science, plan the future, and hike in the mountains.

In 2011, I convinced Nathan to move his lab of more than twenty graduate students and postdocs to ISB. It wasn't long before he was a full professor and associate director of ISB. With the move, he was able to focus all his efforts on his main interests: systems medicine and, increasingly, the development of scientific wellness.

Nathan had come to believe that we almost always wait to deal with disease until it's far too late. Even when someone proactively goes to their doctor with early warning signals, a typical response is, "Come back once you've had this terrible symptom, and at that point, you'll get a drug to help you manage that symptom." We all know someone who has had an experience like this. But Nathan had also had friends who were able to dramatically improve the quality of their lives by getting more information about what was going on in their bodies and by taking action to alter the course of their health. He came to believe we could do much better and that it was vital that we move beyond disease-focused health care to a more active managing of wellness.

Nathan and I started talking about what more could be done to optimize wellness and began exploring ways science and technology could drive better outcomes and allow people to take more control of their own health. In time, as we spoke and collaborated, we became convinced of the need to gather dense data over long periods of time, beginning with healthy people—something that was clearly lacking in contemporary healthcare. This was a passion we shared, and we loved working together. So we partnered on this effort in earnest in 2013, and in 2018, I convinced Nathan we could best move our vision forward by merging our two lab groups to form the Hood-Price Integrated Lab in Systems Biomedicine.

In 2016, Nathan received the Grace A. Goldsmith Award (given to one researcher each year under the age of 50) for his work pioneering scientific wellness. Three years later, he was named one of ten "emerging leaders" in health and medicine by the National Academy of Medicine. In 2020, when he poured himself into understanding the effects of the microbiome on human aging, he received a Catalyst Award in Healthy Longevity from the National Academy of Medicine, and in 2021, he was appointed as one of eighteen members of the influential Board on Life Sciences for the National Academies of Sciences, Engineering, and Medicine. In the fall of 2020, as the coronavirus pandemic was still raging and Pfizer announced the success of the world's first mRNA vaccine, Nathan became CEO of Onegevity, a health artificial intelligence startup, and, following a merger and IPO, chief scientific officer of Thorne HealthTech, which is dedicated to bringing much of this vision of scientific wellness into practice.

Together, Nathan and I have published more than 1,100 peer-reviewed papers in many of the world's top journals, helping to inform global scientific understanding of genetics, molecular immunology, genetic expression, Alzheimer's disease, cancer, metabolism, aging, and health data. I communicate all this not only because I want you to see in Nathan the brilliance that I have seen but also because I want you to appreciate what he is putting on the line by challenging the status quo.

Scientific Wellness

In 2014, Nathan and I put together a pilot project to explore the potential of using genome and phenome analyses to quantify states of wellness and disease in 108 individuals, mostly friends, over a period of nine months.[24] We called this group the Pioneer 100. There were two spectacular results from this project. The first was the realization that we could strikingly improve individual health trajectories with this approach. Over just a period of a few months, we could optimize wellness and, for some, meaningfully delay or even reverse disease transitions. The second was that the data we generated could be analyzed to identify new approaches to treatment and new medical insights. In a study of just a little over 100 patients, we had collected enough information to keep many researchers busy for many years.

As we pored over the data, we began considering what we could do by growing the experiment from 100 patients to 1,000, 10,000, and beyond. We started feeling that this could be the beginning of a big shift, a paradigm-changing Big Sky moment for healthcare.

Over the next days and weeks, we spoke frequently and fervently about this potential moment of inflection. Could the quantification of health and data-based healthcare optimize the human health experience? Could we help usher in the widespread adoption of the principles of scientific or quantitative wellness? Could we define a new kind of twenty-first-century healthcare—and figure out a way to make it available to everyone?

We think so.

This book captures some of the discussions and discoveries we have shared since the time we recognized this Big Sky moment. In the pages that follow, we will describe this coming shift to wellness and prevention, a transition so profound that the medicine we practice today may soon seem like something from the Dark Ages. We do not use these words lightly. The period we're about to enter promises to be transformative. This change, we believe, will take place over the span of one to two decades. In many ways, it has already begun. Scientific wellness offers us an opportunity to reject the idea that chronic diseases are just part of life. There is no reason for us to accept that heart disease, diabetes, cancer, autoimmune diseases, Alzheimer's, and other chronic illnesses are an inevitable part of the human experience. These conditions can be caught long before clinical symptoms make it evident that they exist. They can be stopped early. And if prevention is no longer an option, when the diseases are caught early enough, most can be reversed.

Scientific wellness has the potential to uncouple age and aging, giving us a chance to live into our nineties while continuing to be mentally alert and physically active. It can help us keep our minds as healthy as our bodies, for it truly does not matter what we do to sustain ourselves if our brains do not continue to function effectively. This must be a revolution of health for both the body *and* the mind.

This book is a guide for what is to come. We will explain how medicine will be transformed by optimizing wellness and preventing the flare-up of most chronic diseases so that we will not have to cure them. We'll show that these changes are long overdue and describe the reasons why, despite great resistance and the immense complexity of the challenges, we are so convinced that this revolution is inevitable.

We'll describe an approach centered around patients and a data-driven wellness that will soon (finally) replace the disease-dependent approach that has dominated healthcare for more than a century, reversing disease long before you *feel* sick or have overt symptoms. We'll address the complexities of human diseases and explain why we are finally at a place where we have the power to intervene at just the right time. We'll describe a path that will give you a better chance of not being part of a generation born *just too early* to take advantage of this fundamental shift to human wellness. We will tell you how to begin to optimize your health today and offer insights into how to age in a healthy manner so that you will move into your eighties and nineties physically agile and mentally capable, no matter what the world throws at you.

In short, we are on the cusp of a fundamental shift in the way we think about healthcare—turning it into a system that is truly, finally, and honestly aimed at maximizing our individual well-being. As in all good things, there is a catch. This is *participatory* healthcare. That means that people must actively participate in making choices that will maximize their own health. Society can take steps to make this easier, and healthcare needs to support these efforts. It will be important that access to these choices be available to everyone. But in the end, each of us must become the pilot for guiding and optimizing our own wellness.

Book Organization

Scientists should be judged not by their bold predictions but by their records of achievement. There is an important caveat to this rule, however: those who are too afraid to dream big visions will not achieve big goals. To aim high you must be willing to accept failure. We have strived to embrace big visions and bring them to fruition—and we believe scientific wellness has the potential to revolutionize our conception of what is possible to achieve for health.

In Chapter 1, we explain how the twentieth-century healthcare model came to be, why it was so successful, and why it isn't right for the challenges we now face. Chapter 2 considers how scientific wellness went from an idea to the guiding principle of a healthcare revolution. In Chapter 3, we share what we learned from applying scientific wellness to thousands of people. In Chapter 4, we look at the ways the necessary data of scientific wellness can and should be collected and implemented going forward. Chapter 5 looks at how the relatively new concept of biological aging fits into the healthcare future we envision and how slowing aging will play a central role in delaying the onset of most chronic diseases. Chapter 6 centers on why long-term brain health has long been ignored by our healthcare system—and how we can turn this around. In Chapter 7, Lee shares a story from his own life about Alzheimer's disease—a very personal story that up to this point he hasn't shared with many people outside of his closest friends and family members.

In Chapter 8, we explain how the principles discussed in this book appear to be offering a breakthrough in Alzheimer's research—and why we believe that these successes are yielding a model for fighting other chronic diseases. In Chapter 9, we delve into the exciting work that is happening now in the quest to end cancers—and why we think there may be a huge leap forward coming in this long fight. In Chapter 10, we discuss why artificial intelligence is so important to the realization of these and other healthcare breakthroughs, and in Chapter 11, we conclude by mapping out the path we believe must be taken to make the shift from disease to wellness and prevention a reality.

What Lies Ahead

Today, we have the ability to measure billions of pieces of information on every human being in the world through genome and phenome analyses. We can sample hundreds of different biological systems in each individual—the immune system, hormonal systems, nervous systems, metabolic systems, and more. In doing so, we are unraveling mysteries that, just a few years ago, were so complicated we didn't even know there was a mystery to be solved, and we are arriving at deep insights into the pathology of human wellness, transition, and disease.

We are gaining more insight into what a state of wellness actually looks like and the extent to which it can be optimized. Through cross-disciplinary biology, research scientists are developing a spectrum of measurement, imaging, and computational tools that will help them dive even deeper into the complexities of human biology and disease. All of these advancements are giving us a chance to remake healthcare in a revolutionary way—to end our reliance on a model based on disease and replace it with one that prioritizes wellness from the start. Our aim with this book is to begin to chart the way forward to a time when doctors will treat disease before it is clinically recognized as disease and spend a lot more of their time keeping us well.

The several key paradigm shifts that brought us here followed quite logically from the preceding one, although in no case was it possible to see the next until the last had materialized. This movement, from shift to shift, sometimes reminds me of the endlessly undulating Beartooth Mountains of Montana where, until you reach the next ridgeline, you cannot imagine what type of terrain might lie ahead and what the obstacles might be. And so you explore a promising path. Sometimes, you have to double back. But then you set your sights on the next peak and continue to move forward. In a place so beautiful, so magical, and so mysterious, what else would you want to do?

We hope you'll enjoy the climb, and join us in our enthusiasm for what we can see just over the horizon.

1

An Infectious Idea

Why Healthcare Designed for the Twentieth Century Doesn't Work Anymore

Consider the United States in the year 1900—a time when every third coffin was filled with the body of a child under the age of 5, and the three leading causes of death were pneumonia, tuberculosis, and diarrhea.[1] Now consider your own family and do the morbid math. What is unthinkable today was simply the way things were just over a century ago, as infectious diseases—far deadlier than our recent struggle against COVID-19—ran rampant in a rapidly urbanizing nation. This was a world without vaccines or antibiotics, and without basic responsive medical care for conditions that today can be cured effectively. This was a world of doctors without answers.

What little help there was to combat infectious disease outbreaks was coming not from physicians but from efforts to increase sanitation and hygiene, like the construction of public water supplies and waste-disposal systems. Doctors were often relegated to standing by, helpless when it came to diagnosing and treating even the most common killers of the day. Imagine what it would have felt like to bring your child to a doctor and to hear that doctor say, "We cannot help her." This was the state of medicine at the turn of the twentieth century.

It took an outsider to shake the American medical system to its core and release it from its wretched helplessness. Abraham Flexner wasn't a doctor or a researcher. He was a teacher who had gained a reputation for the disruptive educational methods practiced at the preparatory academy he founded in his hometown of Louisville, Kentucky. At "Mr. Flexner's School," as it was called, there were no grades or examinations, no standard curriculum, and no academic records. "It was essentially a large-scale experiment in individual tutoring," historian Thomas Neville Bonner wrote.[2] And yet Flexner's graduates were making their way to elite universities faster than those from the far more famous preparatory schools of the day. Flexner's fame spread quickly. His style brings to mind, and may have helped inspire, the modern trope of the maverick educators who risk it all to teach their students in ways that buck convention.

Flexner parlayed the growing interest in his unusual methods into an opportunity to study, at the age of 40, at Harvard University, where he began work on a critique of the lecture-based educational system, which he faulted for making teaching expedient for researchers, who didn't really want to be in the classroom, while doing very little to actually teach students. When it was published a few years later in 1908, Flexner's *The American College* gained the attention of Henry Pritchett, the president of the Carnegie Foundation, who asked Flexner to turn his critical eye to medical schools.

It was an odd choice. Flexner had no medical experience, although his brother Simon had been a pathologist at Johns Hopkins University Medical School and the University of Pennsylvania before being appointed the first director of the Rockefeller Institute for Medical Research. That relationship may have given Flexner a modicum of understanding of what he would find when he swung open the doors of American and Canadian medical schools, but it could not have prepared him for what he witnessed once his inquiry had begun.

In collaboration with Johns Hopkins's founding dean, William Welch, Canadian physician William Osler, and Baptist minister Frederick Gates, Flexner set to work, reading everything he could on medical school operation and governance before crisscrossing North America in a whirlwind tour of more than 150 medical schools. To his dismay, he found an untrained, disorganized, often for-profit system of education that was poorly preparing most doctors for the realities of providing medical care in an age of infectious diseases. Most graduates had no idea how to deal effectively with pneumonia or influenza, and Flexner found that many schools were training students in treatments that were not driven by rigorous evidence. Admissions standards were haphazard. Most medical schools were disconnected from hospitals, rendering much of their instruction theoretical. Laboratories, when they existed at all, were poorly equipped with even the basic tools needed to identify the most common diseases, relegating diagnoses to hunches and superstitions more than science.

"In Salem, . . . I asked the dean of the medical school whether the school possessed a physiological laboratory," Flexner wrote in his autobiography. "He replied, 'Surely, I have it upstairs; I will bring it to you.' He went up and brought down a small sphygmograph—an instrument designed to register the movement of the pulse."[3]

That was it. That was the lab. It was a far cry from our modern experience of sprawling academic medical centers with thousands of researchers and a dizzying array of advanced equipment.

Released in 1910, *The Flexner Report* pulled no punches. At Washington University in St. Louis, Flexner wrote, he had found "an ambitious and substantial institution" whose medical department was "entirely out of harmony with the spirit and equipment of the rest of the science-driven university. Unless this department is to be a drag and a reproach, one of two courses must be adopted: the department must be either abolished or reorganized."[4]

This censure was typical of the rest of the report, which called for medical schools to incorporate evidence-based care tied to rigorous scientific studies and clinical trials in an effort to bring science to the practice of medicine, disease diagnosis, and therapy discovery and teach this standard of medicine to students. We are now so used to knowing that scientific studies and clinical trials support what doctors recommend that most of us take it for granted. But it wasn't always so; the modern biomedical scientific enterprise is actually quite new.

To arrive at this new model for medical education, Flexner recommended greater collaboration with medical facilities and stronger regulation of medical licenses. Backed by tens of millions of dollars in philanthropic support for institutions that embraced these reforms—a whole lot of money back then—the report quickly transformed medical education in the United States, leading to the elimination of for-profit schools and establishing new gold standards in medical practice and education: a biomedical model keenly focused on putting scientifically trained physicians front and center in the fight against infectious diseases.

In the century after Flexner's report, medicine grew into the massive enterprise we know today, with evidence-based protocols driving physician training and patient treatment. Gone are the days when medicine was built on superstition, and treatments were mostly ineffective or outright harmful. Led by a new army of physicians trained in the principles of biology, physiology, and biochemistry, twentieth-century medicine brought us a world of antibiotics, vaccines, and sanitation measures that were incredibly successful at eliminating the major infectious causes of early death in the developed world. Polio and smallpox were essentially eradicated. Chicken pox, measles, and pneumonia are no longer the mass killers they once were.

We are not yet—nor will we be any time soon—at a place when all life-threatening infectious diseases are eradicated, as COVID-19 reminded us. The rise of antibiotic-resistant strains of killers like tuberculosis is a significant challenge, and infectious diseases remain a deadly obstacle for the developing world, where typhoid, malaria, AIDS, and tuberculosis continue to kill approximately three million people every year. For this reason, the paradigm of twentieth-century medicine—a "find it and fix it" approach, with therapies focused on targeting and neutralizing a single causal agent—remains very important. But that paradigm has proven ill-suited to tackle the complex chronic diseases that have come to be major killers today—both directly, as in the case of diabetes, cardiovascular disease, obesity, cancer, and neurodegenerative diseases, and indirectly, as comorbidities make infectious diseases such as COVID-19 far more deadly.

The Problem with "Find It and Fix It"

Back in the days when infectious diseases were far more rampant, particularly among the young, it would have been quite ridiculous to worry about problems that, for the most part, hit people in their sixties, seventies, and eighties. The prevalence of tuberculosis, diarrhea, and pneumonia on death certificates naturally informed how the field of medicine was structured following Flexner's damning report. The critical goal for twentieth-century medicine was to find the problem and fix it—period. So it was that infectious diseases informed the development of healthcare for the rest of the century, with systems focused on single-disease causes. What resulted was a vast reduction or even elimination of many of the known invading agents that are distinct from the human body.

When this "find it and fix it" approach was translated into the treatment of chronic diseases, it manifested itself as an intense focus on single-drug therapies. This focus is problematic. Single-drug therapies do not work for most chronic diseases for the simple reason that chronic illnesses are not distinct from the human body and are thus far more complex to treat than infectious diseases. Many involve aberrant mutations of healthy tissues, like cancers, or autoimmune problems, in which the body's own defense system starts to malfunction and attack healthy cells.

Single-drug interventions can help alleviate suffering in *some* cases. Take the drug adalimumab, commonly sold under the brand name Humira. Millions of people around the world with rheumatoid arthritis, ankylosing spondylitis, Crohn's disease, plaque psoriasis, and ulcerative colitis can attest to the fact that their lives are better because of this "blockbuster drug." No one will claim that the drug *cured* them, though, because that's not what Humira does. As with most chronic diseases for which there are available pharmaceuticals, the drug is targeted at the symptoms of the disease, not the causes. Humira inhibits the activity of an overactive immune system, with varying degrees of effectiveness depending on people's genetic predispositions and distinct biology.

But in nearly a third of patients, this drug does little or nothing at all—because many patients' immune systems create antibodies that prevent it from working.[5] And even when the drug does work to alleviate symptoms, it's no panacea. The benefits come along with side effects, weakening the immune system for serious infections such as hepatitis B as well as sometimes triggering allergic reactions, nervous system problems, and heart failure. If you stop using it, the benefits quickly disappear. Humira has been wildly successful commercially, with more than $150 billion in global sales through 2021.

Now, put yourself in the shoes of a drug developer: if you can come up with a single pill that helps reduce the symptoms of a common chronic disease—potentially alleviating the suffering of a lot of people around the world, even if it's not a cure—why wouldn't you do so? Making a small difference is certainly better than making no difference at all. And in the case of drugs aimed at the symptoms of chronic illnesses, making a small difference can also result in a big fortune. It is thus no wonder that so much time and effort has been spent in the pursuit of single-drug solutions to widespread chronic illness. Increasingly, however, it is becoming clear that the pursuit itself is often fruitless, not only failing to find a cure but also falling far short of meaningful symptom reduction.

Perhaps the best example is the long history of failed drug trials for Alzheimer's disease, a decades-long series of disappointments. There have been hundreds of such failures in the quest to slow, stop, and reverse Alzheimer's, nearly all following a similar path, as drug companies and patients put their faith in a process designed to identify a single agent, not the solution to a complex combination of interwoven causes. After all the time, effort, and money spent in this pursuit, only a handful of drugs have been approved for Alzheimer's treatment. This might not seem bad until you consider how little these drugs do. As the Alzheimer's Association puts it, "while these drugs may temporarily help with symptoms, they do not treat the underlying causes of Alzheimer's or slow its progression."[6]

What does "temporarily help with symptoms" mean? In most cases, it means a very small reprieve over a matter of months. The truth is these drugs really don't do much at all. They do cost a lot of money, though, and they come with a host of serious side effects.

Consider what it is like to bring a loved one to a doctor and have that doctor say, "We might be able to make her a little more comfortable, but I cannot cure the disease." All too often, this is the state of medicine in the first decades of the twenty-first century (Figure 1.1).

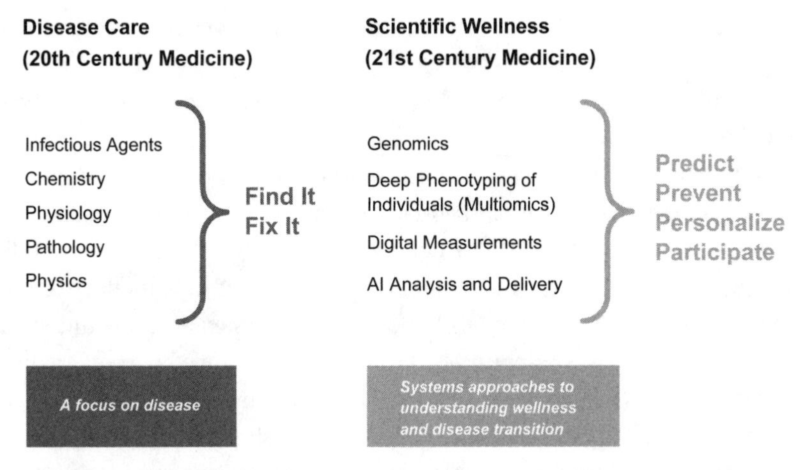

Figure 1.1. Scientific wellness shifts the focus of healthcare from treating late-stage disease to earlier and more personalized interventions.

Alzheimer's is not an outlier. Cancers, diabetes, severe mental illnesses, and autoimmune diseases such as rheumatoid arthritis, lupus, and multiple sclerosis are among the many diseases with multifactorial causes in complex human systems that we have not yet solved. We cannot fight these diseases with one-size-fits-most single "miracle molecules." That's been clear for decades, yet pharmaceutical companies keep trying. They fail far more often than they succeed, and they pay for every new failure by raising the prices on those drugs that work.

"Work" is a relative term. Differences larger than one standard deviation between pharmaceuticals and placebos—meaning a shift large enough to move the average outcome for treatment beyond 34 percent of untreated people in a normal distribution—are uncommon.[7] Most drugs only work for some people, to some extent—improving life marginally, and sometimes not even that much once the side effects have been taken into account. Cancer physicians often tell us they would not choose to go through the chemotherapy regimens that some of their patients do, especially in cases where the benefit is marginal and the discomfort from the treatment extreme. In the case of the most life-threatening cancers, many say they would join an experimental trial focused on the newly emerging science of immunotherapy. Of course they would; they are in the know, and not just when it comes to cancer.

When a joint team of researchers from Germany, Austria, and the United States came together in the mid-2010s to examine the effectiveness of many of the world's most common medicines, they concluded that physicians probably needed "a more realistic view of drug efficacy." By way of example, they pointed to antihypertensive drugs, such as thiazide diuretics, ACE inhibitors, and calcium channel blockers that reduce blood pressure by minuscule amounts, and also noted that the response difference between aspirin and a placebo for the prevention of cardiovascular events was a paltry 0.07 percent per year.[8] When Nathan gave a talk to a group of physicians in 2019 and showed data on how ineffective the top ten drugs sold in the United States were, a physician in the audience suggested in anonymous comments that this "must have been made up." It wasn't. The statistics came from quantitative medicine expert Nicholas Schork, who published the data in the prestigious journal *Nature*.[9] Figure 1.2, also from the *Nature* article, visually captures the astounding fact often kept from so many of the millions of people dutifully taking multiple medications a day.

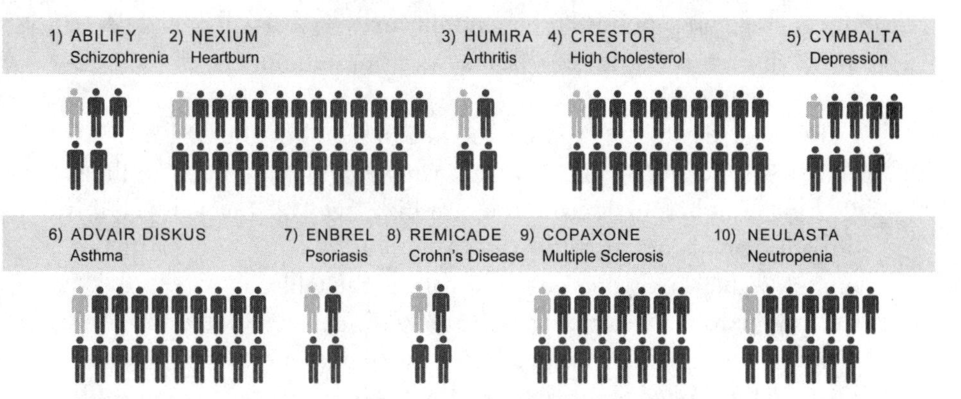

Figure 1.2. At best, the ten highest-grossing drugs prescribed in the United States help no more than one in four patients. Some help as few as one in twenty-five.
Reformatted by permission of Springer Nature from Nicholas Schork, "Personalized Medicine: Time for One-Person Trials," *Nature* 520 (2015): 610.

The drugs represented are Abilify for schizophrenia, Nexium for heartburn, Humira for arthritis, Crestor for high cholesterol, Cymbalta for depression, Advair Diskus for asthma, Enbrel for psoriasis, Remicade for Crohn's disease, Copaxone for multiple sclerosis, and Neulasta for neutropenia. Among these pharmaceutical best-sellers, the best result is that one in four patients benefit (with Humira, Enbrel, and Remicade). The worst response is a benefit for just one in twenty-five patients (with Nexium). Overall, fewer than 10 percent of the patients benefited, and they were all exposed to the undesirable side effects of most of these drugs.

Americans spend $600 billion a year on pharmaceuticals. Because we're still pitifully bad at knowing which drugs will work for which person at which time, much of that investment is wasted on drugs that do not work but do come with serious side effects. The good news is there is promise on the horizon. There are major public initiatives to increase the information we have for personalization, including the NIH's All of Us program, which is sequencing a million people and creating banked samples available for researchers to do all sorts of follow-on measurements. The UK biobank has already generated massive data for genomics and associated other data types for 500,000 people. For cancers, precision medicine using genomic targeting of therapy selection is being made broadly available via companies such as Tempus and

Foundation Medicine. Additionally, Lee is launching a new million-person project based on genomic and phenomic analyses that will provide deep data to distinguish drug responders from non-responders, as we'll discuss in Chapter 11.

The medicines we use to treat the health scourges of the twenty-first century simply don't work that well, yet we continue to pour billions into long-odds bets to discover the next big drug. Poker players call this "throwing good money after bad." When good players see someone doing this, they know they've got a "fish" at the table.

We're the fish in this analogy. Every last one of us.

More than 86 percent of total healthcare dollars are spent on chronic disease in the United States, where $1 out of every $5 spent is on healthcare.[10] That's $4 trillion, rising at more than 6 percent per year—a rate that will be impossible to fund if it continues for even fifteen years.

You'll likely agree that $4 trillion is a monumental sum. Still, it could be money well spent if we were getting tremendous value from all of that cash. But while healthcare spending has gone through the stratosphere, the benefits have not. Analyses have shown that excessive health spending creates economic activity but diminishes national productivity.[11] Rising costs without commensurate benefits create a tremendous economic drag, as chronic health conditions lower both our productivity and our well-being.

As a society, we've struggled to grasp the real problem. When US politicians debate healthcare, the debate is almost entirely focused on how to get everybody insured, with progressives generally arguing for more government-driven remedies and conservatives usually fighting for private solutions. Both sides miss the point: while "coverage for all" is a laudable goal, it doesn't really matter how you pay for something that is broken if you have no plan to fix it. And our system is badly broken—not simply because it doesn't cover everyone but because even those who are covered are getting a product built for a bygone era. The quality of our individual healthcare is far below what it should be.

Four Trillion Reasons Not to Change the Status Quo

Abraham Flexner's report was published in 1910, seven years after the Wright Brothers first took off from Kitty Hawk. Just think how much aviation has changed since then. Today's passenger planes may rely on the same basic principles of aeronautics that helped the Wrights take off from a sand dune in North Carolina, but few would argue that a modern Boeing 777, which has an operating range of up to 8,555 miles, is basically the same thing as the Flyer I, which traveled 852 feet on its longest flight. Would you get on a plane whose design had not meaningfully changed since 1910?

By most estimates, the United States has stalled in extending average life spans.[12] We may, in fact, be moving backward. It's now well established that the COVID-19 pandemic sent average life expectancy plummeting, falling a year and a half in the first year of the epidemic, according to the Centers for Disease Control and Prevention—though among Blacks and Hispanics, the number fell by three years.[13] But even before the coronavirus struck, US residents were in the middle of the longest sustained life span decline since World War I.[14] The same is true for our health spans: nearly 70 percent of Americans between the ages of 55 and 64 suffer from diabetes, cardiovascular disease, chronic obstructive pulmonary disease, asthma, cancer, or arthritis. Among those over the age of 65, more than half are carrying two or more such conditions.[15]

Almost everyone acquires more such conditions as they age, making life difficult, painful, and frankly more unlivable, to the point that many of us have had the surreal experience of feeling an intense sense of relief when a loved one dies.[16] "At least her suffering is over," we say. And while it's difficult to admit, there's often another form of relief for those of us who have gone through the horrible experience of paying for medical care that is doing little to alleviate a loved one's suffering—and might actually be prolonging their agony.

What a terrible reflection on the state of our healthcare. Can our current system be adapted to the new disease paradigm? That's doubtful—and it probably won't take you four trillion guesses to figure out why. Four trillion dollars a year is a lot of money. And although the way it's being spent isn't good for most peoples' health, it has contributed to profitability for some medically oriented constituencies.

As with any industry, a lot of people in the medical establishment and pharmaceutical industry are invested in the status quo, including doctors and hospital administrators, drug developers and insurance companies, medical equipment manufacturers, hospital owners, and politicians whose reelection campaign accounts are supported by one or all of the above. And this isn't just the case in the United States, where private insurance is king, but also in nations with nationalized medical care—where the government *itself* is deeply invested in the way things are, with large armies of bureaucrats whose livelihoods depend on operating the system, not disrupting it.

It should come as no surprise that stakeholders are resistant, or even hostile, to change. That doesn't make them evil. We need to remember that the system we have is the product of decades upon decades of success. It still saves lives. But we've been allowing the people who are profiting from healthcare to define health in a way that reminds us of a scene in *The Good Place*, where guileless amateur DJ Jason Mendoza tells Chidi Anagonye how to play "special Jacksonville-style pool."

"You hit whatever ball you want, but you use your hands. And that's it," Jason says, as he crashes the balls into one another. "OK, your turn. I got 1,000 points! Oh, that's the other rule. You make up your own points."[17]

Modern healthcare is a big, expensive game of Jacksonville-style pool, where the people who have the most to gain are the ones who define the rules. This is a world in which emergency transportation to a hospital can cost more than $10,000, a hospital stay can cost $100,000, drugs that work only slightly better than a placebo can cost thousands or even hundreds of thousands of dollars, and a cutting-edge procedure can cost $1 million, even if none of them end up significantly improving a patient's life.[18]

As it stands, two of the three parts of the triangular relation comprising insurers, healthcare providers, and patients stand to gain when healthcare costs go up, and the market incentives are actually greater—much greater, in fact—when people stay sick than when they are healthy or get better. If you can keep someone alive, even if they are in agony, you can keep billing them. For a deep view into the economics of today's healthcare, we recommend *The Price We Pay* by Johns Hopkins professor Dr. Marty Makary.

The cost of healthcare is bankrupting families and forcing people to make hard decisions—even those who do have health insurance. Donna Talla, who contracted COVID-19 before vaccines were available, was charged $150,000 for her life-saving care despite having private insurance. She found herself faced with cruel choices. "I think I'm going to have to sell my house in order to pay these medical bills," the 56-year-old resident of Virginia told *The BMJ*.[19] "It wasn't part of my retirement plan, but if I have to do that then I will."

This is nonsensical, if not downright immoral, but it makes perverse sense in the context of how healthcare is currently structured (to fight diseases once they have fully emerged) and paid for (as reimbursement for treatment of specific diseases). This puts all of us involved in healthcare in the position of valuing added years of sick life but not added years of healthy life. By not focusing on prevention, it is as if our healthcare system places no value whatsoever on a healthy life.

Innovations aimed at keeping people well are generally only adopted if they can save money for an already profitable death industry and if those savings can be realized in the short term. There is no comparable requirement that treating disease with a drug will cost nothing. We have faced this resistance time and again as we've worked to help large healthcare organizations understand and implement our vision of scientific wellness. This challenge is surmountable, as we will discuss in the coming chapters, but it nonetheless remains a major hurdle. In most cases, if these innovations cost anything at all in the near term, even as an investment in long-term health outcomes, they are almost always dismissed outright. William Sage, an authority on health law and policy who is both an attorney and a surgeon, noted that "The last several decades of medical innovation have mainly involved reimbursable technologies that fit existing, flawed methods of production and therefore that have tended to increase costs without dramatically improving health outcomes."[20]

Put simply, under the current healthcare system, wellness isn't profitable. That's a problem that is further complicated in the United States by the fact that individuals whose coverage is tied to their work—about 58 percent of nonelderly people—are pushed from insurer to insurer and healthcare provider to healthcare provider every few years when they change jobs or their employer selects a new coverage plan for all employees.[21] In a world in which no provider can be assured that the investments they make in a patient's health will be rewarded with savings down the road, there's not much incentive to make that investment. No wonder wellness has largely been ignored in favor of a one-cause-one-solution paradigm that all too often ignores prevention.

Twenty years after sequencing the first human genome, we still haven't harnessed our individual genetic codes' ability to help us identify the unique and diverse ways we are susceptible to diseases. This means that we are not preventing those disease transitions from happening when we could. Healthcare today generally waits for visible symptoms to treat illness, allowing conditions that might otherwise be successfully treated early to grow to the point that treatment is either ineffective, terribly difficult, or prohibitively expensive. And it makes little attempt to educate patients about their own responsibilities when it comes to optimizing their well-being—something doctors are almost never trained to do.

By way of example, although nutrition is critical to health, typical medical training involves the equivalent of a single day of instruction in nutrition. About 10 percent of medical schools don't teach doctors about nutrition at all.[22] "It's a scandal," David Eisenberg, an adjunct associate professor of nutrition at Harvard University's T. H. Chan School of Public Health, told PBS in 2017. "It's outrageous. It's obscene."[23] But, outside of pockets such as functional or integrative medicine, that doesn't mean anything has changed.

The lack of training in nutrition mirrors that in other important areas of human health, including exercise (more than half of the physicians trained in the United States received no formal education in physical activity), sleep (medical school students across several nations receive about three hours of education on healthy sleep), and stress (just 3 percent of visits to a primary care physician include any discussion on this topic).[24] The same is true of brain health, where missed and delayed diagnoses are common in part because physicians mistakenly believe cognitive health and problems like Alzheimer's are not an issue for patients who are not elderly.[25]

It's not just wellness that is sidelined by a healthcare system whose profit incentives are almost fully based on treating diseases. Disease care suffers too. Although there is no question that medical outcomes can be improved with emerging technologies, studies have long shown that healthcare is one of the slowest technology adopters across all industries.[26] Imagine if the phone or computer you purchased today were no faster than it was ten years ago, even though you paid far more for it. You'd be outraged.

We should all be outraged by the sorry state of many of the inefficient and misaligned healthcare structures that prevent us from living longer, healthier, happier lives. A system that has done tremendous good in the past—and still does some good today—is ill adapted to the health challenges of the modern world, namely the complexities of chronic diseases. We find ourselves in a situation similar to the one that confronted Flexner in the early 1900s. It took a shock to the system to change medicine then. We need another shock now—an even bigger one.

Day by day, more and more doctors are coming to see the need for such a revolution. Despite the risks of destabilizing an industry in which the average primary care physician earns about $250,000 a year and the average specialist takes in around $350,000, doctors are some of the strongest supporters of the need for change.[27] The question is how to recruit those advocates to a sustained effort to transform medicine, a challenge we will return to in Chapter 11.

For all the suffering it caused and social dislocation it brought about, the COVID-19 pandemic may be the catalyst that helps bring together doctors, patients, policy makers, and healthcare providers to change the predominant medical model from one that is based on responding to diseases to one that is focused on ensuring sustained wellness. This terrible time in our history has given us ample evidence of the ways wellness can save lives and money. Those who fought COVID-19 from a position of good health were significantly less likely to become seriously sick or to die of COVID-19. There were exceptions, of course, but by and large COVID-19 brought to light the stark differences in susceptibility between patients with chronic conditions like obesity, diabetes, lung disease, and heart problems and those who maintained an optimal body weight and were generally healthy. The pandemic further demonstrated the power of new medical technologies, from mRNA vaccines to telemedicine.[28] Perhaps most strikingly, it gave nearly everyone a glimpse of the frailty of the current system, as physically and emotionally depleted doctors and nurses were sometimes left to care for patients in hallways, parking garages, and tents while major health insurers reported their biggest profits ever.[29]

The Unimaginable Future

There was a time in the history of our species—not so long ago, in fact—when you could have a pretty good idea what your future would look like by looking at the life your parents had led. Farmers begat farmers.

Traders begat traders. The movement from one stage to the next was so slow as to be almost imperceptible.

That is no longer the case. These days it's hard to see even a few years into the future. When Bill Gates was writing *The Road Ahead* in 1993, he famously downplayed the possibility that the internet, as it was constructed, could ever lure mainstream consumers. Between the time he wrote those words and the time the book hit the shelves in 1995, Gates had experienced a change of understanding, though it came too late to stop the release of the book he'd just written.

Back in 1998, Lee was asked by a drug company to predict where medicine would be in fifty years. Knowing that you cannot know where you are going without knowing where you have been, Lee went back to the textbooks and papers of 1948 and found that most biologists at that time believed the genetic material of life was . . . protein. Not DNA, which is made of nucleotides, but the nitrogenous compounds made up of large molecules in chains of amino acids that are key structural and functional components of organic tissue. Imagine how far off the mark we would have been if we'd made a fifty-year prediction about where biology (and medicine) was headed without knowing about DNA!

The same is true of 2076, just over half a century away. The tricentennial of the United States of America is a good marker for envisioning the future of healthcare: it's close enough that many of us will see it, but far enough away to give us space to let our minds explore the possibilities. When we envision this time, we see a world in which the very notion of symptomatic disease is anathema and the biological experience of being 80 is not substantially different from that of being 50. Even the most well informed among us will vastly underestimate, or completely miss, many of the technological developments that will occur between now and then. One need only watch an episode of the original *Star Trek* to see that many of the technologies that Gene Roddenberry imagined would be "cutting edge" in the twenty-third century were quite commonplace by the start of the twenty-first century. We'll overestimate some things, too, but overall, whenever we imagine the future, we tend to *underestimate* the coming changes.

All this is to say that if something we predict seems fanciful right now, the potential we are describing might be old news by 2076. Much of it might be old news by 2036. Our hope is that it will be.

The ultimate goal of helping us achieve lives largely free of disease and the discomforts of aging is obtainable if we are willing to work for it. The participatory role individuals must play in taking charge of their own health will be critical. This is a future in which the health of every person—from fetal development through infancy and childhood to adulthood—can be optimized to the point that most of the diseases that hobble our lives today will be vestiges of an older, darker past. It may seem far-fetched—and it is a daunting challenge—but this is not a pipe dream. It is already under way, backed by scientific data, and there is no reason to believe life can't look like this if we take the necessary steps.

Ray Kurzweil's "Law of Accelerating Returns" speaks to technological progress, but you would have to live in a cave (deep underground, without internet access) not to recognize that medicine is an informational science. In 1950, medical knowledge was on pace to double every fifty years. By 1980, the doubling time was seven years. By 2010, it was cut to three and a half years.[30] And the rate of growth continues to increase. There were 153 exabytes (2^{60} bytes) of global healthcare data generated in 2013 alone, which rose to an estimated 2,314 exabytes generated in 2020.[31] This acceleration is happening irrespective of how all that information is being used. Although there is a vast difference between gathering information and implementing it, it is clear that over the next fifty years we'll experience a massive increase in medical knowledge. It's fun to speculate about the future—and we have to dream if we want to achieve anything of great significance—but this book isn't just about what will happen in the decades to come. It's about our options today and what we should be doing now.

So where should we start? We propose eleven guiding principles for our vision of twenty-first-century medicine.

Healthcare Should Be Wellness Centered

The healthcare of the future will be just that: *health*care. That may be what we call it today, but it's not what it actually is. What we have right now is really sickcare, where the majority of our medical efforts and the overwhelming majority of our spending are devoted to caring for people *after* they get sick, with a huge proportion spent on our worst years. While our last twelve months may make up one-eightieth of our life span or less, that final year gobbles up one-eighth of all the money spent per person on healthcare in the United States.[32]

This is insane. For a beautiful and deep treatise on the problems with end-of-life healthcare, we recommend Atul Gawande's book *Being Mortal*, which makes the compelling case that we should focus less on postponing the end and more on making the rest of our lives as healthy and meaningful as possible. Healthcare should target the maintenance and enhancement of wellness. It should prioritize preventing diseases over fighting them once they've arrived. It should center on the principle of keeping us in tip-top condition to live healthy lives for the long run.

To get there, we need health monitoring networks that are fully informed about the various interlocking biological systems in our bodies. Resulting insights can offer us a chance to get ahead of diseases long before clinical symptoms manifest. We may think of doctors as great physicians when they can spot symptoms using the tiniest of clinical clues and their developed intuition to guide the diagnoses. But in the future, that paradigm will be turned on its head. Doctors who allow their patients to even *get* symptomatic should feel as if they have failed.

The measurements of wellness we should be collecting—the genome, phenome, and digital measures of health—are far more detailed and subtle than "how we feel." Together, they capture information on hundreds of different biological systems. If we begin collecting these measures in a state of wellness, the resulting data can predict wellness-to-disease transitions that are imperceptible to our conscious minds. These transition measurements offer us a chance to be earlier, faster, more precise, and less disruptive in taking corrective actions—and far more effective at stopping disease before it becomes irreversible.

Healthcare Should Be Informed by Your Genome

Sequencing the first human genome was a massive undertaking that required the skills and expertise of hundreds of scientists armed with the world's first automated DNA sequencer, the 370A from Applied Biosystems, which came out of Lee's lab at Caltech. But these days, genome sequencing has become routine, and you can have your genome sequenced for a few hundred dollars in just a few days.

Your genome provides information about the components of the many systems in your body. It is totally unique, and as we learn more about how to read it and pull out what is relevant to individual health, it will form a blueprint for *real* personalized medicine that leads to cures for genetic diseases and provides insights into how to keep you healthy throughout a long life.

As genomic information becomes more widely accessible, we will have the power to redesign healthcare to focus on the specific needs of individuals rather than what seems to work for the people who happen to have been enrolled in a clinical trial. Clinical measures from your blood, such as LDL or HDL cholesterol, should be interpreted through the lens of your genome. How your genetic risk profiles manifest in the body can give you a personalized blueprint for avoiding disease and insight into how genetics can predict which health choices are likely to have the most success.[33]

Beyond using a person's genome to better inform health decisions, it is possible that genetic codes could be manipulated to enhance health, though this won't come without considerable controversy. When He Jiankui used the CRISPR technique to alter the genetic code of two human babies, it triggered an international scandal, and rightfully so. It was much too soon to take this dramatic step, but the field is progressing rapidly, and we will soon be able to identify and correct single-gene genetic diseases before birth, making inherited diseases such as Huntington's or cystic fibrosis a thing of the past. Even now, by using *in vitro* fertilization, one can avoid passing on known dominant genetic diseases, ending the scourge of debilitating inheritances passed through families for generations. Among those who have taken this step is one of our collaborators, Jeff Carroll, who shared how an early diagnosis of Huntington's disease impacted his family planning decisions in the moving documentary *Do You Really Want to Know?*[34]

Your Healthcare Should Be Informed by Your Lifestyle Choices and Life History

Your genome does not define you. Even if we were to create a genetic clone—a person with the exact same genome—that person would not *be* you. Not even close. To realize this, you have only to interact with any set of identical twins. Even right after birth, when their life experiences are almost exactly the same, their personalities, reactions, and interests begin to diverge, and these differences become more pronounced over time. One becomes obese, one becomes an intense endurance athlete. One gets diabetes, one gets into a car accident. One develops a lifelong yoga practice, one becomes a weight lifter. One focuses on nutrition and sleep, one likes to party.

Each person has a unique life experience that leads to a present state of health, including their mental health, and all of this is influenced by a unique genome, temperament, life choices, and environmental exposures. This personal history affects your health and leaves signs that can be read in many ways, including through the mix of biochemicals in your body or blood and the molecular markers that make up your epigenome—chemical modifications that alter how your genetic code gets expressed. Evaluating these signs, many of which we are just coming to understand, will enable the healthcare of the future to be individually tailored in ways that are almost unimaginable today.

Healthcare Should Become More Data-Rich

Human beings are incredibly complex, and the amount of data needed to understand the functioning of our bodies—let alone identify and reverse diseases at the earliest stages—is staggering. When you understand this, it's hard not to feel a bit queasy about the way medicine is currently practiced. Most doctors base their diagnoses on a very small number of lab tests and clinical observations. The kind of "precision prevention" we will discuss in this book is rarely even attempted. There are practical reasons for this: the costs are high, the available information limited, and many doctors have grave concerns about false positives. As anyone who has ever taken a statistics class knows, it can be challenging to wade into even one simple data set.

Consider the data doctors have been working with for centuries. Their patients' heart rates can rarely drop below 35 beats a minute or exceed 200, but there's tremendous variation between the two, including a large span of "healthy" heart rates. Systolic blood pressure can fall well below 60 and exceed 200. Respiration can range from fewer than ten breaths per minute to greater than twenty-five. Currently, doctors routinely test for dozens of actionable biomarkers, screening for lipid levels, thyroid hormone, HbA1c, and fasting glucose, but there are thousands more that could be tested. Every new test creates more variables, and increasing the number of such variables being considered creates virtually endless combinations. This is not like a conventional game of chess; it's more like three-dimensional chess, or *n*-dimensional chess, with pieces that change over time. And that's *before* we get to the human genome. We've just begun factoring that three-billion-base-pair code into medicine, providing a more complete picture of our molecular selves, which should inform our interpretation of the component data.

Modern technologies make possible the collection of massive amounts of relevant physiological health data from digital devices. This data harvesting starts with very common "wearables" like a Fitbit or Apple Watch, which measure steps, elevation gained, quality of sleep, pulse, oxygenation, cortisol levels, and other indicators continuously. Newer wearables will eventually measure proteins, metabolites, mRNA transcripts, and so forth—the essential tools of systems biology and the substrates of the data-intensive medicine of the future. Sensor networks are expanding everywhere now, and healthcare will follow suit with continuous monitoring of blood, urine, feces, saliva, voice recordings, and imaging. We won't just collect information once in a while; we'll collect it continuously.

As "sensor networks" are deployed ubiquitously in the decades ahead, more and more health-relevant data will be passively gathered at an unprecedented scale. Some may not want to participate in this wellness opportunity, and that should remain their choice. But as is the case with those who allow their driving habits to be remotely monitored by insurance companies, there may be incentives to participate, such as decreased healthcare costs (even today, some insurance programs decrease insurance rates for nonsmokers).

Is it possible for healthcare to keep up? We believe it is. But to make that happen, we are going to need help and a lot of education, so that people can make informed choices and society and industries can be organized to be more supportive for health. We'll need to harness all the latest tools of computational clouds, big data search, digital health data management, and hyperscale artificial intelligence (AI).

Healthcare Should Be Systems-Driven and Powered by Artificial Intelligence

Take the smartest doctor in the world. Clone her. Clone her again. Clone her a thousand times over. It will not be enough to manage the explosion of data that will emerge from twenty-first-century medicine. The vast amount of data needed for this vision of healthcare will be beyond the comprehension of any human and beyond the processing power of any group of human brains without significant assistance. AI will be essential to help interpret these data and deliver insights to physicians.

Just as doctors have always searched for patterns in the smaller data sets with which they have been working, AI will seek patterns in the infinitely more complicated data sets that exist now and in the future. And if there is one thing AI is good at, it's finding patterns. Mining the data will take a lot of processing power, but once a potential correlation between data and disease is recognized in one person, it can be searched for in another—and it wouldn't take a supercomputer so much as the equivalent of an internet search to find the same pattern in still others. The more the data-to-disease connection is affirmed, the higher the search can be prioritized in larger population groups. From that point, processing needs will increase again, but our capacity for dealing with data—both in terms of computation and storage—has been on the exponential rise for many decades, and it will continue into the foreseeable future.

AI is already being used by Harvard Medical School and other health systems to provide prediagnostic recommendations to doctors and patients and to double-check diagnoses against data to help reduce costly, and sometimes deadly, medical errors. In the future, AI will not only churn out observed relationships gleaned from massive data sets that may be uninterpretable to humans, it will organize the data systematically into hierarchies of functional networks that span the body: from molecules to cells, from cells to tissues, from tissues to interconnected organ systems, from organ systems to the whole body, and ultimately from the whole body to different subsets of the population, allowing us to tailor medicine by ethnic groups and beyond—all the way to personalization.

Those systems don't stop at one person's body. AI can evaluate genomic groups, social networks, and ecosystems, bringing together data and knowledge to help lead us all to a much healthier future.

Healthcare Should Aim to Eliminate Chronic Disease

We believe the future of medicine will give us a chance to eliminate the symptomatic stages of chronic diseases that we now accept as unavoidable facts of life. That's a bold statement, to be sure, but not a fanciful one. The Chan Zuckerberg Initiative (established by pediatrician Priscilla Chan and her husband Mark Zuckerberg) has focused a good deal of its considerable wealth on exactly this mission of curing, managing, or preventing all disease by the year 2100.

Chan, who is famously rational, certainly doesn't believe it to be a crazy gambit. "If you think about it, penicillin didn't exist eighty years ago," she told CBS journalist Norah O'Donnell in 2019.[35] The discovery—and impact—of antibiotics has hardly been linear. Likewise, the discovery of biomarkers and drug targets will explode as more individual data are accumulated for both wellness and disease conditions.

While it's easy to point out that the efforts to stop even a single disease, like Alzheimer's, have been fraught with failure, there are reasons to believe that a paradigm shift is under way. Take cancers: the advent of personalized immunotherapies, which boost the body's natural defenses against cancer cells based on patient-specific biology, have given experts like Karol Sikora, the former chief of the World Health Organization's cancer program, reason to believe we are on the cusp of a "revolution" that may lead to a massive reduction in life-threatening cancers through the use of custom-tailored therapies.[36]

Cancers are just one class of chronic disease. What of heart disease, stroke, diabetes, obesity, and chronic lung disease? What of autoimmune diseases, even those that may not be deadly but make life harder and more painful, like arthritis or lupus? These are very different conditions, with very different wellness-to-disease pathologies. The tie that binds them, and nearly all diseases, is that they are accentuated by aging, which brings us to our next principle.

Healthcare Should Be Focused on Healthy Aging

Ask anyone if they want to live to 120, and they're likely to tell you that it sounds awful. Ask them if they'd like to live that long if they could be free of pain, mentally alert, and physically agile, and they're likely to reconsider.

Medicine is worthless if it extends our lives but leaves us in pain, barely able to move or think clearly for decades on end. In fact, it's worse than useless: it's inhuman. Nathan recalls the horrible final years of his paternal grandfather, whose only words, forced out in a difficult whisper, were, "I want to die." Family members watched, waited, and prayed for mercy as they awaited a "natural" end. But there is nothing natural about this sort of end. For most of human history, once a debilitating disease struck, death would come quickly. Now we prolong it endlessly. This isn't just backward; it's cruel.

Hand in hand with the goal of enhancing wellness and eliminating chronic diseases, our healthcare must be redirected from extending suffering to extending health. At every point in the process, it should be aimed at enhancing a person's health span, the portion of life that is lived free of disease and chronic pain. That's where our health efforts and resources can be put to their best effect.

What's the ideal health span? Think not in terms of years, but percentages. Whatever age we live to, our health spans should be as close to 100 percent of our life spans as possible. Increasing the percentage of time we live healthy lives quite logically increases the likelihood of living longer, so we can have both longer *and* healthier lives.

There's an important distinction to make here: age and aging are not the same thing. Your chronological age is what you earn with every turn around the sun. Aging is an accumulation of conditions—from gray hair and sore joints to reduced cognitive ability and physical deterioration—that make life harder along the way. You can't slow down the planet's orbit. You can, however, slow aging. People who eat and sleep better, exercise more, stress less, and avoid environmental toxins experience fewer of the conditions of aging as they grow older. In fact, human studies are already providing glimpses into the possibility of reversing the aging process, allowing us to become biologically younger. Even if one cannot reverse the aging process, it is clear that it can be substantially slowed. For most people, slowing the effects of aging would offer a greatly improved chance of a health span that extends into their nineties. The result can be more years—many more, in fact—of mental alertness, physical activity, creative productivity, and fruitful well-being.

All of these factors are why age itself will ultimately no longer be a satisfactory indicator of general health, as it will be increasingly hard to tell the difference between 20 and 40, or 40 and 60. How far this can be pushed is a topic of intense debate in the scientific community, but there is little question that a push to slow aging is coming.[37] In fact, it is already under way. The question of who will get to benefit from these changes will be one of the great moral decisions of human history, as access to health span–extending treatments and therapies have the potential to either greatly reduce or greatly aggravate entrenched disparities. The social, economic, psychological, and political implications of such a transformation can hardly be overstated.

To get to that point, we're going to have to get better at tracking biological aging. We also must learn to more accurately assess the aging of vital organs such as the brain, immune systems, liver, and kidney, optimizing all of these parts of our bodies both independently and holistically. We're going to need to learn how to optimize shifting hormonal balances as life stages progress, recognizing different needs at various stages of life, including pre-pregnancy (for mothers and fathers alike), pregnancy, infancy, toddlerhood, adolescence, menopause, and a greatly extended period of adulthood that ends, at some point, with a short period of rapid decline (that is, a total systems failure) and death—nothing like the decades of accumulating morbidity we face today.

Healthcare Should Be Regenerative

Even among the biologically young and exceptionally healthy, there may be times when the parts that keep us functional begin to wear out. Healthy people are still prone to accidents, injuries, and aging, and they may be more susceptible when they live more active lives. If a person tears a ligament or badly damages a lung in a mountain biking accident, or severs an artery while working in a woodshop, are they out of luck? Well, they shouldn't be. That's why another area where we will be seeing breakthroughs—and enormous efforts back it now—is tissue engineering and regenerative medicine.

You may recall seeing an arresting image of what appeared to be a human ear growing on the back of a mouse. The structure was grown by seeding cow cartilage cells into an ear-shaped mold and then implanting it under the mouse's skin, where it continued to grow. That picture came from some of the field's earliest pioneers, Harvard surgeons Joseph Vacanti and Charles Vacanti, back in 1997.[38] The "replacement parts" field has grown a lot since then, as the aging of the Baby Boom generation has resulted in a demand for tissues and organs that isn't being met by donors. Researchers in China reported in 2020 that they had successfully implanted engineered patient-specific ear-shaped cartilage into children suffering from microtia, a congenital external ear malformation.[39] En route to an eventual goal of engineered organs of all sorts, we have already seen tissue-engineered skin being used to treat burn patients and replace skin lost to diabetic foot ulcers, and engineered bladders transplanted into patients whose bladders were damaged by age, accident, or disease. Other lab-created tissues have been used to help repair and regenerate broken bones.[40]

It is entirely plausible that we will eventually be able to grow everything from hearts to lungs to eyes from our own stem cells and have those pieces "ready to go" for surgical replacement when necessary. It may also be possible to regenerate our tissues continuously by replenishing stem cells (bringing them back to the states and quantities that are characteristic of youth) or to regrow damaged tissues through a targeted manipulation of genes that would enable us to return any cell to a state of pluripotency, a reversion back to a state akin to a stem cell, allowing it to redevelop as a healthy cell of any type. Stem cell work has been fraught with setbacks. Progress has been slower than initially promised, but it is now possible to correct a gene defect in a relevant stem cell population and transfer these corrected stem cells back to the donor. This is especially true for white blood cell defects and their correction from engineered hematopoietic stem cells, which are commonly found in bone marrow.

However these advances come to pass, the result will be healthcare that doesn't just focus on keeping our original parts in working order but also ensures we have the replacement parts we need when we need them. Our vision would be incomplete if it did not include a push toward medicine like this.

Healthcare Should Be Focused on Optimizing Brain Function

It does us no good to stay healthy in body if we aren't healthy in mind—a fact readily understood by so many of us who have seen someone suffer from Alzheimer's disease or dementia or acute mental illness. As we continue to improve our capacity to forestall the kinds of diseases that turn our bodies against us, the next frontier will be tackling conditions that destroy our brains. The healthcare of the future will focus on exercising our brains just as we exercise our bodies and testing and refining our cognitive abilities in response to this exercise as we pass through life. The brain is enormously plastic and hence retrainable. If they haven't gone too far, lost or diminished cognitive functions can be detected and improved.

We should soon be able to identify blood biomarkers that allow us to determine the earliest stages of a wellness-to-disease transition in the brain and use a systems approach to identify therapies that will reverse the disease process at its earliest stage. These efforts will involve optimal nutrition, exercises that challenge all parts of the brain, and imaging to see functional changes in dozens of different substructures. These features can be interpreted in the context of digital twins (we'll explain this concept later) to simulate the factors that will lead to maintaining brain health throughout each of our individual lives. All of this will allow us to take corrective actions as soon as it becomes apparent that something is going awry. In the future, it may even be possible to "back up" information stored in our brains and store it elsewhere, just as we do with computers. (And, yes, the implications of such capacities are complex.)

Healthcare Should Mostly Be Practiced in the Home

The days when healthcare is inexorably associated with hospitals and clinics may soon be coming to an end. We're already moving in this direction—a shift made necessary by the unsustainable capital costs of building and operating hospitals, and by the inherent disadvantages of bringing so many sick people together in one place. The pace of these changes was significantly accelerated by the COVID-19 pandemic, which prompted the mass adoption of telemedicine, brought about huge leaps in at-home disease testing, and encouraged many doctors and other medical professionals to return to the age-old practice of making house calls—though this time, virtually.

Enabled by at-home health sensors and personal AI, along with far simpler treatments of disease, healthcare will be fully centered on the individual and family at home, with in- and outpatient treatment reserved for procedures that require highly specialized equipment, post-accident emergency care, and diseases that manage to slip through the cracks.

It is reasonable to ask whether the technology required for those who wish to have their healthcare needs met at home will be available to everyone, regardless of income. The answer must be "yes," and this investment could be a major step toward earning the trust of communities that have been given plenty of reasons to be wary of the current healthcare system. While eight out of ten white Americans say they trust doctors to do what is right "most of the time," a Kaiser Family Foundation survey found that only six out of ten Black adults felt the same way.[41] Ensuring that everyone can fulfill the majority of their healthcare needs from home won't erase health disparities, but it's a start.

Healthcare Costs Should Be Much More Transparent

If you're looking for a reason distrust in our healthcare system is so pervasive, especially among the economically disadvantaged, a quick peek at local bankruptcy filings will help clear things up. Medical expenses contribute to two-thirds of all bankruptcies in the United States.[42]

Bankruptcy, of course, is the process through which individuals seek relief from debt—and there is no shortage of people who scorn the bankrupt as those who are seeking to skirt responsibility for their own bad decisions. But that's not how medical debt happens, because patients often have no idea what they will be charged at the time they arrive at their doctor's office, a specialized clinic, or hospital.

This is an obscenity. Indeed, it should be illegal. The Affordable Care Act has taken some steps to improve this situation, but billing horror stories continue. Motivated by their upset voters, Congress passed the No Surprises Act—which went into effect in 2022—with the hope of improving this situation. Almost all of us have either had such an experience or had loved ones affected. Healthcare today is bloated, inefficient, and a labyrinth of complexity with many intermediaries increasing costs. With more personalized control and automation, we will pay for healthcare—either individually or socially—knowing in advance what we are getting and what it costs.

Healthcare Should Be Predictive, Preventive, Personalized, and Participatory

Taken together, the eleven principles above lead to a healthcare practice that reads signals in the body to predict health changes before they manifest, knows how to intervene and act on these predictions, understands each individual, and allows every person to be the central driver of their own health. This vision of predictive, preventive, personalized, and participatory healthcare would meet the challenges we face now and beyond. The key question is how to operationalize P4 healthcare.

When will we get there? That depends on the decisions we make right now. It doesn't have to take long. There is a great tradition of rapid and radical change in healthcare, especially when it becomes clear that change is long past due, and even more so when patients lead the way. You can play a role in optimizing your own health trajectory by pushing your providers to work with you to put wellness first in your care and/or by finding a physician who is responsive to wellness-oriented care. But just as Abraham Flexner's ideas about medical education would never have come to fruition had his historic report not come with the backing of the philanthropic institutions that embraced his reforms, these principles will not come to fruition without significant government and private support.

That's happening, too. Let us tell you where we are finding it.

2

Catalyzing the Revolution

How Wellness-Centered Healthcare Went from an Eccentric Idea to a Vision for the Future

In 2005, Thomas Friedman published *The World Is Flat: A Brief History of the Twenty-First Century*. At 488 pages, the book was hardly "brief," though the time period it addressed certainly was. At that point, the twenty-first century was in its toddler years, but Friedman argued that the world had already changed in dramatic ways. The internet, he wrote, had made the world flat in the sense that information was now available on a global scale. One could reach across the world for best practices in education, industry, retail, and more. Science and technology enabled this shift, and scientists and technologists have benefited from it. More than five billion people now use the internet—about 65 percent of the global population. Over the last twenty years, healthcare practices have also been "flattened." It is possible to tap the best experts and source materials, and secure the resources needed to attack big problems through integrated strategic partnerships. This is more than a benefit of the modern world. It is a requisite of modern science—crucial to making possible the big revolutions in biology and medicine that are needed to tackle the complex health challenges we now face.

When Lee set up the Institute for Systems Biology (ISB) in 2000 with his cofounders, Swiss proteomics pioneer Ruedi Aebersold and South African immunologist Alan Aderem, he was determined to bring together expertise across disciplines in biology, chemistry, physics, mathematics, and data analytics. One of the first things Lee did was recruit young scientists with exceptional skills in biology, data analysis, and technology. They came from Canada, the United States, the United Kingdom, China, and

Iran. He felt it would be transformative to draw on scientific knowledge being generated all across our newly flattened world.

Over the first five years of its existence, ISB built an exciting multidisciplinary culture, with scientists coming from around the world to carry out pathbreaking studies. We mentioned two of these studies in the Introduction—predictive modeling of the halobacterium's responses to environmental signals and a study of prion neurodegenerative brain disease—but there were many others. Teams of researchers investigated the mechanisms governing immunity in mice and humans, looked into how yeast transports proteins from the cytoplasm to the nucleus, and developed novel computational tools for proteomics. Yet despite our successes, we remained mired in a funding paradigm conceived for small science. Central to the future of systems biology was the need to identify and quantify organ-specific blood proteins and to carry out multiomic analyses of important biological systems (this is what we were after in our investigations of innate immunity). We knew we would have to experiment with new technologies and generate new algorithms to help push forward the exciting new fields of genomics, proteomics, and cell biology. None of this would be cheap.

Our successes slowly began to convince other institutions to establish centers for systems biology across the world. In 2006, Ruedi Aebersold left ISB to set up an outstanding systems biology center at the Swiss Federal Institute of Technology in Zürich, Switzerland. Later, Alan Aderem left ISB to head up the Seattle Biomedical Research Institute, where he introduced systems biology. Both of these institutions contributed to the explosion of interest in systems biology such that today there are more than one hundred centers, institutes, and departments across the world dedicated to systems biology.

The systems approach to medicine brings together teams of scientists working across disciplines aimed at large-scale challenges. By the time Nathan moved to ISB, it was clear that these large, integrated teams would be key to pushing forward systems-based predictive and preventive medicine, or what Lee had started calling P4 medicine. Achieving this ambitious goal would require vision and leadership, both of which we had, but it would also take financial support, which at that time was lacking.

In 2005, Lee set out to identify a technologically advanced, educationally daring nation that might be interested in becoming a world hub for systems biomedicine. His idea was that ISB would help build a systems-driven institute in that country in exchange for the resources to support its P4 research. This infusion of outside funding would be invaluable, as it could be used to establish a "proof of principle" that might help unlock further government and private investments.

After promising but ultimately fruitless explorations with Israel, Ireland, Canada, and South Korea, this quest was starting to look hopeless. But what Cervantes said of Don Quixote was certainly true—*donde una puerta se cierra, otra se abre* (when one door closes, another opens).

In 2007, Gerry McDougall, a close friend of Lee and a partner at PricewaterhouseCoopers, was a member of the team recruited to help the small but wealthy country of Luxembourg diversify its economy. Gerry introduced Lee to Jeannot Krecké, Luxembourg's minister for the economy and foreign trade, a former professional football player who was eager to break Luxembourg's dependence on financial services by moving into healthcare. After a forty-five-minute phone call with Lee, Krecké was fascinated by the possibility of a systems-driven approach to disease. He liked the idea of establishing a new systems biology center in Luxembourg.

Lee asked him how much his tiny nation would invest. "Whatever you need," Krecké pledged. After watching Israel, Ireland, Canada, and South Korea explore and ultimately reject the prospect of collaboration, Roger Perlmutter, the chair of ISB's board, was understandably skeptical. "If this succeeds," he said, "I will believe in the tooth fairy."

ISB's director of strategic partnerships, Diane Isonaka, and her husband, David Galas, an ISB professor, worked together with Lee, a team from ISB and collaborating institutions, Nathan (then a professor at the University of Illinois), and a group from PricewaterhouseCoopers to build a $100 million, five-year proposal to help create the Luxembourg Centre for Systems Biomedicine (LCSB), dedicated to pioneering systems approaches to medicine. Academic inertia had repeatedly stymied advances in systems medicine, but the University of Luxembourg had one big advantage: it was four years old and was only just beginning to recruit its biology faculty. Also key was the fact that Krecké had given the project his stamp of approval. Big new ideas require leadership, determined optimism, and financial resources, and the Luxembourg Centre would never have happened without him.

ISB offered to identify a director and help recruit the faculty. It would train eleven senior scientists and postdoctoral fellows in Seattle in the technologies and computational tools necessary to carry out the research. These scientists would then transfer to LCSB and bring their newly gained knowledge to Luxembourg. ISB also agreed to host exchange visits for LCSB faculty and alternate hosting symposia.

The proposal presented a framework for the creation of an international scientific and technological network that would include scientists from around the world. In exchange, Luxembourg would contribute $100 million to ISB over five years to support the development of a variety of technologies and strategies for P4 medicine, many of which had long been conceptualized but needed funding to get off the ground. This money would also support our collaborative efforts with LCSB. Building the new institute would be funded separately by the Luxembourg government.

It was a bold ask, but Krecké didn't flinch. When the agreement was signed, Perlmutter, good sport that he is, proclaimed his belief in the tooth fairy.

The strategic partnership began in 2008, and by 2011, LCSB was open and completely functional. Rudi Balling, an outstanding geneticist who had led several large scientific institutions in Germany, became the founding director. Under Balling's leadership, LCSB grew and grew. When the five-year strategic partnership ended in 2013, there was simply no question that the Luxembourg Centre for Systems Biology would survive and thrive. Today, LCSB has more than a dozen research groups and more than 200 employees. It has become one of the most respected systems biology institutions in the world.

Together, ISB and LCSB built cross-institute programs in systems genetics, family genomics, P4 medicine, proteomics, and neurodegeneration that resulted in ground-breaking science. The collaboration catalyzed foundational efforts for P4 medicine. In 2010, Lee's lab published the first whole-genome sequencing of a family in *Science*, directly leading to the identification of the causal gene for a rare disease, Miller's syndrome, in an affected child.[1] Another highlight was Robert Moritz and Ruedi Aebersold's development of a new mass-spectrometry technique, selected in 2012 by *Nature Methods* as technology of the year, to quantify hundreds of proteins with a high degree of sensitivity.[2] Another major advance was the optimization of a single-RNA molecule analysis technique by Lee's group, leading to the creation of the company NanoString, which used the technique to detect panels of cancer biomarkers.[3] Lee's lab also helped optimize the inkjet technology approach in rapid DNA synthesis that Agilent still uses today to synthesize DNA fragments and DNA arrays.[4] Jim Heath, then at Caltech and now president of ISB, pioneered a dynamic new technique that allowed peptides to be designed as drugs with powerful specificity.[5]

One of the early projects we both took on aimed to investigate the pathway of neurodegenerative disease through a class of infectious, self-reproducing proteins known as prions.[6] The function of prions isn't clear, but their association with a group of rare neurodegenerative diseases, including Creutzfeldt-Jakob disease, Gerstmann-Sträussler-Scheinker syndrome, and fatal familial insomnia, has been well established.[7] Prions exist in two states—a normal, folded state called cellular PrP^{C} and an abnormal folded state called infectious PrP^{Sc}. When injected into the nervous systems of mice, the abnormally folded state has the ability to transform the normal state, causing more abnormal proteins and leading to disease and death within about twenty weeks.

This course of events may seem far removed from the human experience, but the knowledge is powerful because it tells us exactly when the wellness-to-disease transition was initiated. What that means is that we could compare normal and infected mice to determine how their gene expression patterns differ and, since we know cells use signaling pathways to connect parts of the body to the whole, we could search for clues that something has gone wrong in the specific areas of the brain we know to be impacted by abnormal prions.

We introduced the infectious PrP^{Sc} protein into several hundred mice and followed the twenty-week course of the disease, examining the mice's brains roughly every two weeks. We did the same with 100 uninfected mice, examining their brains at the same time to be sure we wouldn't mistake signs of natural aging for signals of neurodegenerative disease. In particular, we were looking for the ways single-stranded molecules of messenger RNA, or mRNA, differed; we found about 330 such molecules.

About a third of the differentially expressed mRNAs could not easily be associated with a single biological network, and these are still being actively studied. But the other two-thirds were a more immediate gold mine of information because they fell into one of four different biological networks related to this disease: the accumulation and replication of PrP, the activation of microglial and astrocyte cells (two key types of brain cells that interact with neurons), the degeneration of the synaptic connections of neuronal cells, and the death of those neurons.

Not surprisingly, all four networks became increasingly perturbed as the disease progressed, and they became disease-perturbed in an ordered manner. The perceptible disease transitions appeared in the network order listed above, and their successive disease perturbations were roughly four weeks apart, correlating quite well with the advent of pathological signs of the disease in the brain—and affirming the mechanistic relevance of this disease-relevant networked ordering. It's important to note that clinical signs—those that would be used to diagnose disease today—didn't show up until about eighteen weeks after infection. That was long after three, if not all four, of the networks were clearly disease-perturbed. At that point, the mice were about two weeks from dying of this disease. Even more important, we found that several molecules in the blood were changed quantitatively at the earliest stages of the disease progression, meaning that we then had a way to detect this transition using a simple blood draw early on, when the disease is far less complex.

A second project that came out of the Luxembourg partnership was a genetic analysis of bipolar disorder in humans.[8] BD, as it's often called, is a psychiatric condition characterized by episodes of mania and depression that affects as many as one in every thirty-five people in the United States.[9] Because it is so common, BD has been well researched relative to many other psychiatric conditions. Twin and family studies have shown it to be one of the most heritable mental illnesses. Still, its exact genetic basis has remained elusive.[10]

To pinpoint the genetic roots of this disease, we sequenced the genomes of 200 people from forty-one families, each with multiple affected individuals with bipolar disorder, in the hope of identifying rare and common variants contributing to BD genetic risk. We initially focused on 3,087 candidate genes with known synaptic functions identified in prior studies as having an association with BD, and soon homed in on 125 gene variants found more frequently in patients suffering from bipolar disorder. These analyses revealed that many of the gene variants affected neuronal ion channels, the pores through which our nerve cells control the flow of sodium, potassium, and calcium—hence controlling the action potentials that allow nerve cells to transmit information. In the timeworn ways of drug development, this would have been an obvious target—and indeed, many of the drugs that have been approved to treat BD are aimed at these channels—but a systems approach doesn't end there.

Out of those 3,087 candidate genes and 125 common risk variants, we focused on the twenty-five most common variants and went looking for their genetic signatures in more than 1,000 unrelated individuals with BD, an analysis that confirmed the increased frequencies of these high-risk variants. Virtually all of these gene variants were rare, and virtually all of them resided outside the coding regions of their nearest gene. This indicated that they were regulatory in nature and controlled the levels of expression of those corresponding genes. What this likely means is that BD doesn't come about as the result of a single genetic cause, or even a combination of genetic variants acting upon one another, but that there are likely to be many different genetic variant combinations that, in various combinations, may lead to similar pathological outcomes. That is an incredibly important insight to have as we look to the future of the fight against bipolar disorder because it tells us that it is incredibly unlikely that any one treatment or therapy will work for most people, most of the time. Indeed, with dozens of variants at play, it is unlikely that any one intervention will work for more than a small percentage of people living with this disorder. For most, the disease comes from a combination of genetic variants, so a single drug won't be enough. This argues compellingly for the need to detect BD at its earliest stage, as we saw earlier with prion disease. At that point, only one or a few networks will be disease-perturbed—and single drugs or other targeted intervention may actually have a shot at controlling or reversing BD.

That complexity might seem like bad news, but it's quite a promising development. There are literally scores of different BD interventions that have been shown to work for some people, some of the time. The current practice for prescribing these treatments is to take the one that works most often, see if it works, and then move on to another if it doesn't—subjecting patients to a long process of trial and error and exposing them to medications that change their body chemistries in myriad ways. What we can now begin to do is look at what seems to work for individuals with one specific risk variant or a specific combination of risk variants—a mood stabilizer like lamotrigine, for instance, for someone with variant A; an antipsychotic like lurasidone for someone with variant B; transcranial magnetic stimulation for someone with variant C; cognitive behavioral therapy for someone with a combination of variants D, E, and F; and so on. And we can study combinations of drugs for people with multiple variants.

These two approaches—targeted single drugs and combination drugs—are indicative of the sort of highly personalized care we see as central to twenty-first-century medicine. The solution is no panacea—the use of multiple drugs brings increasing risks of side effects—but understanding the disease from a systems point of view makes it possible to tailor treatment to each individual's needs. As more of these types of genome-wide BD variant data are gathered, we will be able to predict risk more effectively and design personalized therapies for hundreds of different genetic variants. And if we can do that for bipolar disorder, which has been so hard to pin down, we should be able to do it for virtually every other genetic disease.

Whether we are speaking of prion diseases, bipolar disorder, or any other medical condition, knowing what disease looks like at the earliest moments of transition is obviously important. But there's an essential component that must be part of these explorations. It's very hard to see a subtle change in something if you don't know what that thing looked like to begin with—and that's why understanding what wellness looks like for each individual is so vital.

Once we figure that out, the next step is to proactively improve people's health before they become sick. And that's no easy strategy.

Measuring Wellness

Largely flying under the radar of medical journalism, some positive things are happening right now in doctors' offices around the world. Research over the past few decades has shown that physicians have long been too reluctant, too busy, or too uninformed to talk to their patients about the importance of sleep, healthy eating, exercise, avoiding toxins, lowering stress, and other essential practices aimed at maintaining better health. That's finally starting to change. Doctors are increasingly making time for these conversations and helping patients incorporate these practices into their lives. The results can be profound.

Many people have come to believe that this is what "wellness-centered" medicine is all about. But that's not quite right, because these conversations almost always begin when a doctor sees that someone is already in a state of transition to disease, if not experiencing clinical symptoms. At that point it's very hard, if not impossible, to turn things around.

This is made even more difficult by the fact that the vast majority of medical research is focused on disease, with very little focused on wellness. Take, for instance, one of the most influential population health studies, the Framingham Heart Study, which for more than seventy years has been monitoring the health status of a large cohort of people, starting in 1948 with 5,209 individuals from a small Massachusetts town. Many of the most important insights we have today about the risk factors for cardiovascular disease and other long-term health outcomes come from this study, which was among the first to employ simple clinical chemistries and physiological measurements related to heart monitoring to create a longitudinal picture of heart health. Principles that might seem like common sense today—like the fact that smoking leads to more than just lung disease, elevated blood pressure increases the risk of stroke, or high LDL cholesterol is a warning sign for heart disease—were established as a result of this study.[11] We are deeply indebted to the thousands of original subjects of the study and the four subsequent generations of Framingham patients who have followed in their footsteps, not to mention the researchers who have kept this effort going for so many years, creating a model for population health studies that has been followed many times since.

While Framingham has taught us a lot about how to do a population study focused on disease, there have been few similarly large and long-term research efforts aimed at wellness. Should there have been? Certainly. Could there still be? Sure. But longitudinal studies are tremendously expensive, and the results can take decades to materialize. So even though we are now coming to recognize the importance of understanding baseline wellness—at both the individual and population-wide levels—it would take us a while to follow this strategy to catch up to where we need to be. Fortunately, we live in an era in which new types of phenomic measurements are being discovered all the time, and we actually have the computational power to process, assess, and effectively interpret the data that come from those measures. If you were to ask a modern supercomputer to crunch every data set and calculate every equation ever used in the thousands of peer-reviewed studies built upon the Framingham data over the past half century, the programming could be done in weeks. And the processing? That would take only a few minutes.

One thing we still can't do is speed up time. But rather than look at hundreds of variables over decades for a few thousand individuals, we can now look at billions of variables over a shorter time span for tens of thousands, hundreds of thousands, and soon even millions of people. The good news is that we don't have to wait for disease to become clinically observable to do this. We can assess wellness in people's early adult years, before most disease transitions begin.

In the early 2000s no one was doing this systematically.

There had been a few intriguing initial dives into deep phenotyping (i.e., making lots of phenomic measurements to clinically identify characteristics that signal health or disease) as leading scientists made a huge number of measurements on one subject over time and published the results. At Stanford University, geneticist Michael Snyder began carrying out the first deep longitudinal profiling of a single person—himself—in the early 2010s.[12] At the same time, a few hours south at the San Diego Supercomputer Center, physicist Larry Smarr was working toward the goal of measuring his gut microbiome health.[13]

"Have you ever figured how information-rich your stool is?" Smarr asked a reporter from *The Atlantic* in 2012.[14] "There are about 100 billion bacteria per gram. Each bacterium has DNA whose length is typically one to 10 megabases—call it 1 million bytes of information. This means human stool has a data capacity of 100,000 terabytes of information stored per gram. That's many orders of magnitude more information density than, say, in a chip in your smartphone or your personal computer. So your stool is far more interesting than a computer."

This is true, but whatever we might learn from Snyder's extensive self-measurements—or Smarr's stool—would only tell us something about *each individual.* To understand whether the findings might be applicable to others, we need a much wider data set. And that's why, in 2013, we launched the Pioneer 100 Wellness program.

The Pioneer 100

The effort began with 108 people.[15] We asked them to commit to a nine-month period during which we would examine their every molecular nook and cranny. First, we determined the complete genome sequence of each participant, offering a baseline of genetic potential both positive (looking at so-called "longevity genes," for instance) and negative (identifying mutations that indicate a greater propensity for cancer, heart disease, diabetes, or Alzheimer's disease). Barely a decade had passed since the world's first human genome had been completed, but sequencing the genome was now the *easy* part. Its interpretation, even today, remains a challenge.

Next, we moved on to the phenome: biomarkers that are influenced by a person's genome, lifestyle (things like diet, exercise, sleep, and stress), and environmental exposures such as radiation, toxins, metal poisons like lead and mercury, and molds. We tested for 1,200 blood analytes—proteins, metabolites, and other clinically relevant chemicals—a test we repeated every three months during the study period. The tests allowed us to assess hundreds of different biological networks. We also took saliva samples to assess the levels of the stress hormone cortisol and examined the gut microbiome with regular stool samples, giving us a picture of the presence and abundance of different bacteria and allowing us to see how they changed over the course of the study period. Finally, we used wearable trackers made by Fitbit to monitor each participant's activity level, pulse, and sleep. We also regularly checked for blood pressure, waist circumference, weight, and body mass index.

During the course of the trial, the Pioneer 100 participants became the world's most measured people. This was a historic feat in and of itself. But measurements alone weren't enough. We wanted to see the measurements change in a direction that indicated increasing wellness. That's why we used each subject's initial phenome, along with what we knew about their base genome, to identify a long list of actionable possibilities for each individual validated by the clinical literature. Every recommendation was chosen either to improve wellness or to avoid or mitigate disease.

Some of these interventions were quite simple. For example, an assessment of some participants' microbiomes showed a near-perfect mix of bacteria for the creation of trimethylamine (TMA), a molecule generated in the gut that gets converted by our livers into trimethylamine N-oxide (TMAO), which has been linked to elevated risks of cancer, cardiovascular disease, and diabetes.[16] Not too far down the line, one could imagine using an analysis of gut microbiota to proactively identify those who are in danger of having high TMAO levels, then prescribing a targeted mix of bacteriophages (which kill specific kinds of bacteria), probiotics, and prebiotics to eliminate and replace the problem bacteria, circumventing the capacity for the gut microbiome to make TMA. Research suggests that simple dietary changes, like limiting the consumption of red meat, can also reduce TMAO production. So that's what we counseled Pioneer 100 participants with this particular deleterious mix of bacteria in their guts to do.

In other cases, we identified participants whose genome indicated an elevated risk for diabetes. That could be particularly troubling if they had elevated glucose levels in their blood, which is a sign of prediabetes. But why wait for that stage of the wellness-to-disease transition before taking action? Research indicates that reducing the intake of simple sugars and increasing fiber in the diet is a good remedy for lowering blood sugar levels, so that was one of the recommendations given to individuals with an elevated genomic risk for diabetes—even if they had no clinical signs of disease.

Now, it might seem obvious that consuming less sugar and eating more fiber is a good plan for just about anyone. It is. Ditto for reducing red meat. So is getting plenty of aerobic exercise, drinking adequate water, and ensuring a healthy level of vitamin D by stepping into the sun for a few more minutes each day. All of these recommendations are good for most people, most of the time. It's hard, though, for anyone to make a lot of changes at once. So what the Pioneer 100 participants got was essentially a personalized checklist that allowed them to prioritize the steps that would offer them the greatest benefits in terms of maintaining or improving wellness.

You might think that people who chose to enroll in a project like the Pioneer 100 would be motivated to follow our recommendations, and in large part they were. But any doctor or researcher will tell you people are woefully bad at following instructions.[17] Even when a physician tells a patient to "take two aspirin and call me in the morning," what happens between that moment and the next morning is anyone's guess. That's why we held education sessions on how the genome, phenome, gut microbiome, and digital health measurements would be converted into a list of actionable possibilities for each individual—and why we were so fortunate to have Sandi Kaplan on our team. As the wellness coach for the Pioneer 100 study, Sandi leveraged her training in psychology and nutrition to persuade the pioneers to change their lifestyles in accordance with the individual recommendations. With expertise, passion, and warmth, she taught them the power and scope of the actions they had been advised to take. In the end, we estimate that nearly 75 percent of the specific actions Sandi recommended were acted on by the pioneers, an incredibly high success rate for such programs.

There was plenty for Sandi and the other team members to help the participants address. The first blood draw told us that 91 percent of the Pioneer 100 had nutritional limitations (such as low omega-3 fatty acids, low iron, or high mercury), 68 percent had inflammatory complications that generally arise from poor diets, 59 percent had cardiovascular complications such as high blood pressure or high cholesterol, 56 percent were prediabetic, and 90 percent had low vitamin D blood levels. Even the healthiest among them had multiple actionable possibilities that, according to clinical research, could result in greater wellness. In other words, they were a fairly typical representation of the American public.

The Evidence Mounts

Nine months is a relatively short window for changing a person's health trajectory. We wondered whether we would see anything meaningful over the course of that period, especially since some of the participants were battling health conditions that were the result of an accumulation of many decades of unhealthy lifestyles. Take Tayloe Washburn, the founding dean of Northeastern University in Seattle. Tayloe was 68 when he entered the Pioneer 100 study. At that point, he had been having arthritis-like symptoms in his knees and ankles for several years, making it impossible for him to do the things he most loved, such as hiking in the mountains with his family. By virtue of his job, he had access to some of the world's best doctors, who had prescribed several medicines intended to treat his condition. But like so many others facing similar chronic conditions, Tayloe wasn't happy with the improvements these drugs offered and felt the results could not really be called "improvements" if you took into account the side effects he was experiencing.

In making their recommendations, his doctors had fallen back on the classic framework for twentieth-century medicine: Take a drug that has been demonstrated to work in some (but nowhere near all) cases, accept that there will likely be some side effects, and wait to see what happens.

Tayloe's doctors based their diagnosis and treatment plan on a small number of indicators about his health, most of which were self-reported and associated with arthritis. They didn't have the benefit of the 360-degree view that comes with complete genome analyses and a deep phenotyping approach. But with hundreds of analytes to review, his genome sequenced, and all of the other data we had collected in the pilot study, a different picture came into view—and it was a game changer.

Tayloe's ferritin levels (a measure of iron in the blood) were significantly higher than what is typically thought to be normal. Now, in many cases this isn't something to worry too much about. It's often just a sign of a diet that is, or has recently been, a little high in meat or other iron-rich foods. Many doctors simply ignore it—and that's what had happened in Tayloe's case. But a review of Tayloe's genome revealed that he had two bad copies of a gene associated with hemochromatosis, a disease in which the body absorbs too much iron. Among the early symptoms? A breakdown of cartilage in some patients that can cause arthritic-like stiffness and pain. By itself, the iron measurement might not have meant much. With this additional information, it was a clue that the problem he really had wasn't the one he was being treated for.

Thankfully, Tayloe's phenome showed no indication of diabetes, liver disease, or heart disease, so he was in Stage 1 of hemochromatosis. At this point, the accumulation of iron in the blood is generally deposited in the joints or ligaments, leading to arthritic symptoms. In Stage 2, the excess iron can also be deposited in the skin and in the pancreas, leading to a bronzing of the skin and destruction of the beta islet cells with incipient type 2 diabetes. In Stage 3, iron is deposited in the liver, leading to fibrosis—and serious liver disease—at which point it can progress to cardiac decompensation and a significant loss of heart function, which can be fatal. Treatment at Stage 1 is both simple and altruistic: give blood periodically until the high iron levels return to normal, and repeat as necessary throughout the rest of your life to maintain a normal blood iron level. Tayloe did just that. He adjusted his diet to avoid iron-rich foods, and it wasn't long before the arthritis disappeared and he was back to hiking in the mountains with his family.

Now, to be fair, it probably shouldn't have taken a 360-degree view of Tayloe's genetic code and phenotype to identify this problem. Iron isn't always included in the common blood panels physicians request, but it's not exactly an obscure blood chemical. Just about anyone who has been through middle-school biology can tell you that iron is vital to the functioning of all cells, and particularly important to red blood cells, as it helps carry oxygen through our bodies. What's more, hemochromatosis is the most common genetic disease among white people, about 10 percent of whom carry at least one of the risk variants, although far fewer actually suffer from it because you generally need two copies of the bad gene—one from each parent—for the disease to be expressed. And yet hemochromatosis wasn't caught in Tayloe's case—a patient who enjoyed superior healthcare, had identified a clear problem, and had every reason to want to get better. Not, that is, until it showed up in the blood work that was being done for the Pioneer 100 study, where we integrated two different types of measurements (genome and blood) to make the correct diagnosis.

We think this is a key point: actionable possibilities often arise from the integration of two or more different types of biological information. These variables are not always available to physicians, and even when they are, some doctors dismiss their potential significance.

Good outliers make for great anecdotes. But Tayloe was not an outlier. In our small sample set of about 100 patients, we found several individuals whose hemochromatosis had gone unchecked.

Another patient whom we will call Jim had been an active blood donor until he was 35, at which point he joined a scientific advisory committee for an organization in Uganda. Because his participation in the organization had potentially exposed him to tropical pathogens, blood donation centers would no longer take his blood. Jim considered this an unfortunate consequence of his service in Africa but, unbeknownst to him, it came at a significant personal cost. Had he continued to give blood after 35, he might have discovered much earlier that he had hemochromatosis—and he would have periodically removed the excess iron from his system to remain completely free of the consequences of that disease. Instead, he began a slow descent from wellness to transition to later-stage disease.

Jim first noted signs of something not being right at the age of 40, when he started losing the hair on his legs. His first doctor diagnosed a bizarre infectious disease, but antibiotic treatment had no effect. Ten years later, Jim noticed what seemed to be inflammation in the tendons of his arms, an effect that left him unable to play tennis, and he began experiencing arthritic symptoms that were so bad he couldn't hold a glass of water without using both hands. Over the following years, these symptoms gradually got worse. Jim went to a series of physicians, none of whom were able to identify the cause of his suffering or fix the problem.

It wasn't until the age of 67, when Jim underwent the deep phenotyping of the Pioneer 100 project, which revealed high levels of blood iron and two defective copies of the hemochromatosis gene, that a correct diagnosis was made. Like Tayloe, Jim's treatment included elective blood draws. This stopped the progression of the disease, and for someone who had spent the past quarter-century getting worse and worse, the halting of the disease's progression was of no small consequence. The treatment regimen didn't reverse all of the symptoms, however. Too much time had passed. That doesn't mean there was nothing he could do—managing any disease is a much easier endeavor when you actually know what the disease is—but the disease had already progressed to an irreversible state, too far to turn back with current medical know-how.

This is an unfortunate story, to be certain. Yet the same impetus that drove Jim to donate blood as a young man and go to Africa in his thirties compelled him to become an early participant in a health revolution. And what we've learned from his experience has already been instrumental in saving others from a similar fate. We don't have to wait until someone is suffering from hemochromatosis to take action if we make the correct measurements early enough. But too often today we fail to do the measurements, and hemochromatosis advances to an irreversible state, as it did with Jim, emphasizing the importance of early detection of a transition and its reversal.

Another participant, Diane, offered us a shining example of how deep phenotyping can catch problems long before they become problems. Diane also had two copies of the high-risk gene for hemochromatosis—a fact she did not know until she joined the pilot study. As a woman in her thirties, she would not be expected to show signs of high blood ferritin levels until she went through menopause. Now she is well prepared. For the rest of her life, she will take the very simple step of monitoring her blood ferritin levels. If they start going up, she'll go to her doctor for an elective blood draw.

Taken together, these examples of three people with the same genetic predisposition to disease—none of whom knew it until they were subject to genomic analyses and deep phenotyping—illustrate the possibilities of scientific wellness and twenty-first-century medicine. There are thousands of other analogous "actionable possibilities" waiting to be discovered by genome/phenome analyses. And you probably have several of these.

Those who come into this new healthcare paradigm in their sixties, seventies, and beyond may not always be able to undo the damage of a lifetime, but they will have a much better chance of stopping the progression of an identified chronic disease. Others may be able to turn back problems before the point of no return, as Tayloe did. But among those who are young or disease free, the opportunities are vast. Like Diane, they will have the information they need to hold off the diseases to which they are most susceptible when they are still in a state of wellness or at a point of very early transition. And if symptomatic disease never materializes—if it never creates pain, causes a loss of ability, or impacts other biological systems in a negative way—what can we call that but a functional end to that disease?

Hemochromatosis is considered a Mendelian recessive disease, which supposedly requires two bad copies of the gene for the disease to manifest. We were able to demonstrate that some individuals have high iron blood levels with just a single copy of the gene, which is why identifying and following all individuals with one or two defective copies of the hemochromatosis gene is advisable.

But hemochromatosis is just one genetic disease. We know of about 7,000 additional Mendelian recessive diseases, including cystic fibrosis, sickle cell anemia, and Tay-Sachs disease. As whole-genome sequencing becomes more common, we will be able to determine whether any given individual is at risk for any one or more of these 7,000 diseases and initiate preventive treatments if they exist.

There are 7,000 or more additional diseases that are generally classified as rare, and about 80 percent of these conditions are simple genetic diseases in which a single defective gene can cause symptomatic disease. Thus, just with the information we have to work with right now, we can potentially identify more than 14,000 different genetic diseases—many of which, when identified early enough, can be thwarted at a stage of wellness or early transition with relatively simple interventions like dietary or other lifestyle changes or a combination of treatments that might include supplements and drugs. Alas, there are plenty of diseases for which early identification isn't enough, but that's in no small part because in most cases we've not been able to intervene before symptoms appear. Now we can and, in the long run, we believe this early knowledge about disease propensity will be an essential element of a twenty-first-century standard of medicine that frees humans from much of the tyranny of the thousands of genetic diseases that lie hidden in our genomes.

The nine months of the study went by quite quickly, but it turns out it was enough time to put to rest the debate over whether our pioneers would see any significant health improvements. By the end of the pilot project, most of the people who had volunteered had experienced striking improvements in their health as measured by a variety of criteria.

Of fifty-three prediabetics, seven returned to normal. Most of the others improved significantly and were reversing their prediabetic state, with dietary and exercise counseling playing a central role in driving a significant shift in average blood glucose levels. The number of participants with out-of-range values for clinical markers of prediabetes dramatically changed: 38 percent improvement for average blood glucose; 19 percent improvement for fasting glucose; and 55 percent improvement for insulin resistance.

The number of out-of-range clinical markers improved for other health issues, too. Over just nine months, there was a 12 percent improvement for the biomarkers associated with inflammation, which is a known driver of many diseases and has been linked to Alzheimer's, cancer, heart disease, and diabetes. There was also a 6 percent improvement for cardiovascular health as measured by the biomarkers associated with coronary artery disease and other related diseases. There was a 21 percent improvement in the markers associated with better nutrition.

The Pioneer 100 program gave us reason to believe in the power of scientific or quantitative wellness. It validated our thesis that scientific wellness could bear dividends. It also instilled in us a deep belief that it was time to offer the transformative opportunities of data-driven health to more people.

A "Mavericky" Idea

The union between healthcare and private money is admittedly problematic. But those who deride the substantial shift, over recent decades, from public to private funding for basic medical research may not recognize that large-scale, government-sponsored life science research is a relatively recent development. Before World War II, health research was primarily sponsored by industry, academic institutions, foundations, and private individuals. In 1940, the US federal government was responsible for just $3 million of the estimated $45 million spent on biomedical research, and researchers were often skeptical of government support.[18] This was a time when Nazi Germany was using the work of medical researchers to further its murderous eugenic aims, so it was not unreasonable for scientists to fear that the government would exert undue influence on their work.

This is not to say that governments should get out of the research game—far from it. Government-sponsored research is a major public good and should be encouraged. Clearly, the Human Genome Project and the $100 million Luxembourg initiative that validated the P4 model have demonstrated that government investment can be critical to the advancement of big science. But these sorts of investments have waxed and waned over time and across the globe. In 2014, as the Pioneer 100 project was exceeding our expectations, we had a strong hunch we would have to seek private support to take the next big steps in bringing scientific wellness to consumers.

This recognition was solidified when we approached the National Institutes of Health to explore the possibility of government funding for a larger wellness project. A conversation with one NIH leader was nothing short of demoralizing. "NIH is interested in disease, not wellness," we were told. "There is no chance we will fund this wellness project." So we turned to venture capital funding.

We were excited to get an early meeting with Maveron, a Seattle-based investment company started by Starbucks founder Howard Schultz and investment banker Dan Levitan, both of whom had excellent reputations for supporting innovative consumer-based ideas. We explained what we were learning about the power of scientific wellness and gave Levitan a vision of the future we saw coming: one in which deep phenotyping wasn't a niche research question but a fundamental part of every person's healthcare. Near the end of our pitch, he made a discreet phone call and, quite quickly, his executive assistant appeared with a manila folder.

"Do you know what Maveron stands for?" Levitan asked us.

We probably should have done our homework, but we didn't have a clue. Thankfully he was unperturbed. He told us that the company name was a portmanteau of the words "maverick" and "vision."

"And what you've just told me," he continued, "was the most maver-icky, visiony story I've ever heard."

He opened the envelope and pulled out a check—for twice what we had asked.

That is how Arivale was born.

The Birth of Arivale

Our new company, Arivale, was based on personalized, data-driven health coaching, with a model very similar to the one at the heart of the Pioneer 100 project. Clayton Lewis, a general partner at Maveron in charge of health investments, became the CEO and third co-founder, with us. Clayton was instrumental to developing the business plan and leading the team.

Most of the management, coaching, and technology teams from the Pioneer 100 program were quick to sign on, so Arivale was up and running almost immediately. We were enrolling clients in no time at all. Participants were paired with health coaches who were registered dietitians, certified nutritionists, and/or registered nurses. These coaches didn't simply tell them what to do; they helped them understand the vast trove of clinical and genetic data available on their personal data dashboard.

"Vast" doesn't really do justice to the scale of the data we collected. The participants' genomes were sequenced at their initial blood draw and their phenomes were sampled every six months using blood analytes, gut microbiome, and digital health measurements with a Fitbit and other devices. Participants' blood and fecal samples were stored in a biobank—a library of personal samples for a future in which we were certain to discover even more analyte- and gut microbe–related insights that would be important for further optimizing wellness. The biobank was also important for testing biological hypotheses that came from the data analyses and for looking for blood biomarkers that marked the earliest detectable point in wellness-to-disease transitions for individuals. It also opened the possibility of searching for relevant new drug targets.

As the participants came to better understand their data, a remarkable thing occurred. We saw a notable and measurable increase in motivation among people who we had thought were already quite motivated. The reason for this surge of positivity quickly became clear.

For years, doctors have told patients to maintain a healthy body weight, which, for most, means losing weight. But weight is just one data point among billions, and one of the hardest to change. The body has built-in mechanisms to fight weight loss. Our bodies evolved when fat was hard won and crucial for survival in lean times, and they give it up grudgingly. Exercise alone, without changing one's diet, is almost never enough to lose weight, and there is no single ideal body weight. Yet we've long been focused on this number—much to our psychological detriment, as anyone who has ever stepped on a scale and felt *emotionally* heavier knows.

The other common numbers we use to measure our health are little better. Blood pressure, heart rate, and cholesterol are products of complex and intertwined physiological processes. These numbers can be hard to move, and their levels vary from person to person.

Arivale created a personal dashboard for each individual that allowed them to follow their relevant phenotypic measurements over time. Now, for the first time in their lives, they were able to watch the numbers shift, improving little by little over time and offering hope, inspiration, and resolve throughout the process of optimizing their wellness.

Life-Changing Results

John, a middle-aged business manager, is a good example of this kind of emotional and physiological shift. He was 100 pounds overweight when he found us, and he told us that he felt sluggish all the time. He wasn't sleeping well. He was prediabetic. It was hard for him to exercise, and he reported low self-esteem arising from his appearance. His work performance was marginal.

John knew his weight was a major problem, but his previous attempts to diet and exercise had always ended in failure. What we provided was an opportunity to focus on something other than the number on the bathroom scale—and the chance to do so with help from a coach who was both knowledgeable and sympathetic to the challenges of establishing healthy habits. By focusing on—and celebrating—small changes as reflected in various individual biomarker levels, John and his coach were able to create momentum toward bigger changes. Energized by lowering the glucose levels in his blood, John set to work trying to change other numbers. Losing weight turned out to be a byproduct of his new resolve. Invigorated by the loss of a few pounds, he began exercising more. And that didn't just help with his weight; it contributed to positive psychological changes as well. That's not surprising, since the association between physical exercise and improved mental health is well established.[19] One unexpected side benefit was that John's increased self-esteem made him a far more effective manager, and his work successes made him feel even better about his overall life.

"For the first time since I can remember," he told us, "I feel good about myself."

Over a period of about a year, with coaching guidance and data input from his genome and blood analytes, John lost more than 100 pounds. And he kept it off—a feat that very few dieters are able to do in the long term.[20]

Another early Arivale patient, Max, also saw his life transformed, albeit in a very different way. For years, he had had a harder and harder time with a palsy or twitching of the muscles on one half of his face. When his coach saw from Max's blood work that he had very low levels of vitamin B_{12}, she was almost certain he was suffering from a type of facial paralysis known as Bell's palsy. At the time, there was no unequivocally known genetic cause for Bell's, which would have been helpful for diagnosing this condition. (Researchers identified the first sequence variant conferring risk of this form of facial paralysis in 2021.)[21] The solution was simple: Max's coach told him to connect with his healthcare provider, who ended up making the diagnosis and treating with injections of intramuscular vitamin B_{12}. Within months, the palsy was gone—and Max was literally smiling like a much younger man.

The important point is that wellness is a multidimensional reality that can be tackled in a variety of ways for each individual. Deep data for each individual offers insights into the many different actionable possibilities unique to that person, then a plan to attack them can be formed in a reasonable, incremental manner that provides immediate reinforcement.

With the sorts of results Arivale helped its clients achieve, most participants found their investment worthwhile. The combination of genome-phenome sequencing, analysis, and wellness coaching wasn't just life-improving—it was life-saving.

Beth's salivary cortisol level was so high her coach had to double-check to make sure there hadn't been an error in measurement. But while high cortisol (tracked dynamically throughout the course of twenty-four hours) is a sign of significant stress in the body, it doesn't offer specific clues as to the underlying cause. What's more, Beth, who was 33, didn't appear to be experiencing the sorts of life events that would lend themselves to tremendous psychological stress.

"What else is happening with you right now?" her coach asked. "Is there anything else going on that could explain why your body is in such a state of alarm?"

Beth thought for a bit before saying that she had recently begun experiencing some gastrointestinal discomfort. But, she hastened to add, it was so mild it was almost not worth mentioning.

"Beth's discomfort might have been very slight," her coach later reflected, "but her cortisol was screaming at us that something was wrong."

At the molecular level, the relationship between stress and cancer isn't completely clear. What is known, however, is that when a person can perceive a problem in their gut, cortisol levels are often found to be high.[22] With this in mind, Beth was encouraged by her coach to explore getting a colonoscopy. When she asked her physician to prescribe one, her concerns were dismissed. The doctor reminded her that she'd had one two years earlier, and it hadn't detected any pathology.

"Go home and forget about all of this," Beth was told. "You're young. You're fine."

As it turns out, the doctor's dismissiveness may have been a lifesaver. If he had spent more time listening to Beth and addressing her concerns but had ultimately declined to recommend the exam, she might have accepted his decision. As it was, she felt disrespected—and that drove her on.

A second doctor was similarly unconvinced a colonoscopy was necessary, but he at least listened to Beth and consented to put the paperwork through for the test. And thank heavens for that. The colonoscopy led to a diagnosis of Stage 3 colon cancer—the last stage prior to metastasis.

Beth was rushed into surgery, where the tumorous masses were removed from the terminal part of her digestive tract. Now, five years later, she is cancer free. How much longer did she have before the cancer spread? That's impossible to say. What is clear is that, if she had followed her first doctor's advice, she certainly would not be alive today.

A Missed Opportunity, a Chance to Learn

The individual outcomes we saw in Arivale clients were validating although not particularly surprising. We learned during the Pioneer 100 program that we could expect these kinds of results. But because Arivale was a business, we needed to find a model that would make scientific wellness sustainable and scalable. We struggled with our balance sheets until Dave Sabey, a data center business owner who has deep insights into running companies and getting the best from his 100-plus employees, offered to fund enrollment for all of his employees. About 80 percent of his staff immediately took him up on the offer, and the obvious benefits to those participating soon persuaded most of the others to enroll as well.

Dave cared deeply for his employees. But he also saw his offer as an investment in his business—one that reaped nearly immediate returns.

"The company's cultural transformation was remarkable," Dave said. "Camaraderie, warmth, concern, improved job performances, and certainly a remarkable increase in the benefits of wellness emerged from the scientific wellness program. The employees reported feeling more well, more alert, and passionately enthusiastic for their job's challenges." What's more, they felt that their boss really cared about them.

One of the greatest expenses for any US company is healthcare. In recent decades, insurance has gotten more and more expensive while workers, by and large, have gotten less and less healthy, making insurance even *more* expensive. But a company that invests in the wellness of its employees isn't just improving morale and effectiveness—it's also cutting long-term health costs, including those associated with workdays lost to sickness, disability pay, and early retirement due to disease.

This is the model that we came to believe would sustain Arivale—a system in which large groups of employees would buy into the culturally transformative and cost-saving benefits of scientific wellness. We felt we got further confirmation that we were on the right track when

Providence St. Joseph Health agreed to sponsor Arivale memberships for one thousand employees in a trial led by Dr. Ora Karp Gordon, an expert in internal medicine and medical genetics.

If we could have built more partnerships like this, perhaps Arivale would have made it. Alas, Arivale failed commercially in the end.

It's rather clichéd to attribute a business failure to the notion that the company was simply "ahead of its time." We'll spare you that sob story. While it might be partially true, one can make the same argument for *any* scientific advance that forces physician practitioners to change their mode of practice. Yet, from time to time, these ideas take hold in big and small ways. Businesses that challenge the status quo don't always survive, but when they do, they have a tremendous opportunity to thrive. Arivale didn't make it.

When Arivale closed its doors in April 2019, the problem seemed clear: "What is tragic on so many levels," CEO Clayton Lewis said in an interview on the day the shutdown was announced, "is that we were not successful in going out and convincing large numbers of new consumers that you could optimize your wellness and avoid disease with a little bit of data and some changes in your lifestyle."

Arivale was seeking to provide a path to a healthier and better life for people who had spent their entire lives interacting with a healthcare system that largely views "not being sick" as the same thing as being healthy. But there is a vast and continuous spectrum from healthy to sick—and many, if not most, people are leaning toward the sick end of the spectrum, even if they don't show any outward signs of disease.

In a sense, each of us is like a car without brakes rolling down a hill. At the moment the car begins rolling, it is still a car, functional in every way a car should be. And if the hill is long and straight, that car can roll for a long time and still look and operate just as a car is supposed to. But if it isn't stopped before a wall, tree, or ditch eventually stops it, that car will very quickly cease being able to do most of the things a car is supposed to do. What we failed to recognize at Arivale is that most people don't yet grasp that they are, in fact, already well down the hill and gaining speed fast. They can't yet absorb the idea that scientific wellness can place them on a healthy trajectory—not just slowing down the decline, but actively turning it around—in part because they don't realize that they are already far from well. This disconnect between people's general perception that there is no health emergency and the demands and cost of the program meant that enrollment was too limited to be sustainable.

The price of all of the measurements we took was a part of this failure. Even when we reduced the price substantially, though, the value of the data was still not clear enough for most potential clients to embrace given their uncertainty about what scientific wellness was and what exactly they were getting. And at the price offered ($99/month at the end), Arivale lost money on any clients who didn't stay for the long run, making the business concept unsustainable. Even in the heavily privatized US healthcare market, most consumers are not used to paying out of pocket for health-related services, and insurance companies don't currently cover the costs of data-driven wellness services. This will change in the coming years, as cost savings and material benefits that can come from wellness and prevention are better understood, but Arivale won't be around to see it.

Arivale was unable to overcome one final challenge: at the time, the Federal Drug Administration (FDA) did not permit genomic companies to deliver disease-related actionable possibilities to their customers without first going through a lengthy regulatory process. In 2013, DNA testing company 23andMe shut down its direct-to-consumer health testing kits for two years after the FDA said the company couldn't provide evidence that each of their many computationally derived tests was individually "analytically or clinically validated."

The obstacles faced by 23andMe and many companies that similarly sought to use new technologies to connect customers with a better understanding of their health were significantly smoothed over under the leadership of FDA commissioner Scott Gottlieb, who served from 2017 to 2019. Today, companies can be evaluated by the quality of the process they use to determine genetic evidence rather than regulatory validation of every individual finding. This major step forward still protects consumers and discourages bad actors but doesn't render the approval process so slow that it cripples innovation.

Our failure to integrate Arivale within existing healthcare systems also meant that we had to hand off findings to our patients' doctors, and the results of those interactions were often less than ideal. This isn't because doctors don't care about their patients—they do—but the vast majority have been trained to look for and treat signs of disease, not to identify and maintain a state of wellness. Very few doctors have been educated on the ways in which genomic, phenomic, and other data can be integrated to derive deep insights into patient health, and as a consequence, many are skeptical of additional testing. And given the already extensive demands on physicians' time, not to mention the number of genuinely harebrained ideas about health that have proliferated on the internet, some doctors view any new idea that comes to them from their patients as a likely waste of their time.

Another major limitation of Arivale is that we did not convince a broad spectrum of consumers or their physicians that scientific wellness will, in time, transform each of our health trajectories to ones that focus on wellness and prevention. The education of patients and physicians should involve the teaching of P4 medicine at the high school, college, graduate school, and senior training levels, and we are developing courses and textbooks to do so. Likewise, physicians need to understand the power of genome and phenome in revealing opportunities to improve each patient's health and to be convinced that medicine is fundamentally about the optimization of wellness for each individual, not the treatment of disease. Ultimately, medical schools should teach courses on precision population health so that a new generation of doctors will understand all that the field offers.

The bottom line is that, even though the health outcomes Arivale offered clients were revolutionary, and its data enabled groundbreaking scientific discoveries, the business didn't make it. We're in good company. The first modern electric car, the EV1, was produced for just four years before being scrapped by General Motors in 1999.[23] That effort set the stage, however, for a small Silicon Valley startup to begin producing luxury electric sports cars. Today that company, Tesla Motors, is just one of many competing for part of a booming global market in passenger plug-in electric cars. In similar fashion, Arivale may offer a model for future efforts to move healthcare toward a wellness-centered system. Lee has cofounded seventeen biotechnology companies, including Amgen and Applied Biosystems (both with multibillion-dollar market caps), yet he considers Arivale his most successful because it clearly illustrated the need for healthcare to move toward a data-driven focus on wellness and prevention.

What needs to happen to ensure the next effort will be more successful? For starters, we're going to need to glean all we can from the wealth of data that was generated and see how it can be harnessed to improve health. Let's head into the data clouds and see how they have informed us.

3

Mining Vast Troves of Information

How Data Clouds Will Revolutionize the Way We Look at Health

To bring a wellness-centered vision of twenty-first-century healthcare to life using the data-driven focus on wellness and prevention that guided our efforts at Arivale, we have to start thinking about patient data in a new way—with "personal data clouds" at the center of everything we do. These clouds will store information about multiple dimensions of each individual's unique and dynamic experience with wellness, transitions, and disease, including their genetic and phenotypic data, measures of digital health, medical history, and more—the same kinds of data generated at the Pioneer 100 initiative and Arivale.

Collecting the data is only a start. Data can be converted into knowledge through computational analysis, but there is not a human brain in the world that can untangle the complexity of any one of these data types—a truth that is made clear by looking at the first data type, the genome. The human genome is composed of roughly three billion nucleotides, the code of A, T, C, and G described earlier. Many of these nucleotides are the same from human to human, but approximately one in every thousand is a variant, creating millions of single-nucleotide variations that make each one of us genetically unique. What's more, the 20,000 genes encoded by our three billion nucleotides don't work in isolation; they function as parts of intricate systems or networks with other genes—and this further complexity places meaningful analysis out of reach of the human mind.

Contrary to popular belief, there is no one inherited gene for eye color. At least fifteen genes—and likely many more—play a role in eye color inheritance.[1] And that's just one relatively simple and unchanging trait. Researchers have identified hundreds of other genes that contribute to other aspects of eye development in mammals.[2] The possible permutations—and patterns hidden within those permutations—are virtually endless and require powerful computational approaches to delineate and decipher.

The Pioneer 100 and Arivale programs didn't stop at the genome. Both endeavored to assess each individual's dynamically changing phenomes, too. Because each one of us is unique, the chemical constituents (what scientists call analytes) that reveal how our lifestyle and environment will interact with the expression of our genes to influence health—captured in proteins, metabolites, daily behaviors via wearables, functional measurements, and so on—can be joined in billions upon billions of different combinations, reflecting the complexities of wellness and disease propensities.

What's more, there are hundreds of species of microbes that reside in our guts in different combinations and quantities for every person on the planet. Any one of these microbial species can impact our health and modify any number of our phenomic readouts. These microbes can turn a benign chemical exposure into a toxin or turn a toxin into something benign. What we ingest can kill off our microbiomes or cause them to proliferate.[3] They can even modify the regulation of our genetic code—prompting a gene that programs our cells to do one thing one day and something entirely different the next.[4]

Add to this the fact that each one of us goes about our lives in different ways. Some of us sleep five hours, others nine. Some of us barely make it to 1,000 steps each day, others take well over 10,000. Some of us sit in a car for hours on our commutes, others stand in a subway car or ride a bike, and still others never leave their homes to work. Some of us don't get much vigorous exercise at all, others run for miles and miles. Some have lives full of stress, while others manage to live peacefully. Some eat a healthy diet, others do not. All of these factors and more are picked up by our digital measures of health—including the sorts of data collected using Fitbit bands worn by the Pioneer 100 and Arivale clients.

If we were to collect a single person's genome just once, do a blood analyte assessment of their phenome a few times each year, conduct a census of their gut microbes every few months, and record their digital measures of health via a simple wearable each day, how many potential permutations (combinations of data) would a few years of measurements give us?

As many as the grains of sand on all the world's beaches? Not even close. As many as the stars in the universe? Nope, that doesn't do it justice either. The number is almost beyond comprehension. And that is just for one person. The good news is that there are now powerful approaches we can employ to begin unraveling the bewildering array of relevant associations to lead to new, actionable possibilities. And begin we must. This is exactly what we believe should be done for each and every individual. And some national health systems are already dipping their toes in the water with plans for widespread sequencing of patients' genomes.

While much remains to be done, we have powerful analytical tools to decipher and make sense of these data. Thanks to machine learning, computer systems can be set to work identifying data patterns that provide clues to the underlying causes of wellness and disease. Computers can be used to assemble the data, translate them into biological networks, and extend these networks to include integrative relationships among very different types of data. These networks can center on processes such as the metabolism of carbohydrates, inflammatory responses to the environment or invading viruses, synaptic activity in the brain, or any of a dizzying array of networks operating in your body this very moment. Right now, all over the world, giant networks of computers—including those stored in clouds hosted by Amazon, Google, Microsoft, major healthcare systems, researchers, governments, and more—are working their way through many different types of individual data clouds and seeking almost imperceptible patterns, as if every member of this army of electronic detectives was dedicated to unraveling the mysteries of each person's body. Few of these efforts are integrating many different data types on the same individuals, an essential operation if we want to define and identify thousands of new "actionable possibilities."

From Arivale clients who agreed to have their health data used anonymously in further research, we now have expansive individual data clouds with multiomic deep phenotyping data for about 5,000 people. (In this case, multiomics, or deep phenotyping, combines genomic information with other data such as proteomics, the study of proteins in the blood, and metabolomics, the study of small molecules such as nutrients in the blood, as well as an expansive set of about 100 clinical lab tests such as your doctor might order. Multiomics can also be expanded to other data types such as transcriptomics, looking at messenger RNAs in white blood cells or other tissues, and epigenetics, to make sense of gene expression and activation, but those types of data collection weren't performed with Arivale.) Some individual clouds extend across four or more years. And with each passing day, their usefulness grows as we analyze and integrate their different data types further, offering new opportunities in clinical medicine, new knowledge for biomarker discovery and therapy development, and new understandings of how our environment, personal choices, and genetics affect our health.

Why Genetic Data Matters

We generally divide human development into a series of stages—fetal, newborn, infant, child, teenager, young adult, middle-aged, and elderly. There are different schools of thought as to how these stages should be classified, but just about everyone agrees that they differ in fundamental ways. The data clouds will provide unique opportunities to explore our biological systems as they travel through these stages and to follow the progression of disease in each of these phases of development.

Although we know relatively little about the molecular transitions that occur as one crosses these stages, what we do know is that, in early childhood and during the teen years and early adulthood, we don't typically have to think much about personal wellness or disease. We generally feel quite healthy during these years. Some of us wish this stage would last longer, and it can. In the young, the rates of cancer, cardiovascular disease, and neurodegenerative disease are so low that when these conditions appear they often come as a shock, even to physicians—and they are sometimes missed until it is tragically too late. But even if a young person is not sufficiently symptomatic to be diagnosed, the earliest hints of some impending diseases are already present. The wellness-to-disease transitions for cardiovascular diseases most often seen clinically in people who are middle-aged and elderly commonly began many decades earlier, as shown in the Framingham Heart Study.[5] And this isn't only true of heart disease. A 2020 study of 2,658 cancers published in *Nature* showed that cancer-causing genetic mutations occur many years, sometimes even decades, before diagnosis.[6] In the brain, PET (positron emission tomography) scans show metabolic deficits as many as ten to fifteen years before cognitive decline sets in for those diagnosed with Alzheimer's.[7]

Imagine if we could identify these earliest transitions and head off diseases long before they clinically manifest. Could we maintain the health and vigor of youth more deeply into our developmental trajectory?

Let's consider the types of actionable possibilities that have arisen from each of two major types of data—the static genome sequence and the dynamically changing phenome. The American Society of Human Genetics has identified seventy-six disease-causing variants for which there are possible actions to prevent disease. Although these variants are generally rare, they are collectively found in about 5 percent of the American population. Just *knowing* they exist and acting appropriately can be the difference between life and death. Consider the following:

- Individuals with the gene variant that causes malignant hyperthermia—perhaps one in every 2,000 people—frequently experience a rapid rise in body temperature and severe muscle contractions leading to death if they are given an inappropriate anesthetic. For these people, a simple operation can be lethal. And yet most hospitals don't screen for this gene before putting a patient under anesthesia. If everyone had their genome sequenced, we could eliminate this problem by determining who had this variant and choosing anesthetics that would not trigger this deadly reaction.
- Variants in two other well-known genes, BRCA1 and BRCA2, can cause breast and/or ovarian cancer. Actress Angelina Jolie boldly and controversially chose to have breast and ovarian ablative surgeries after learning she was a carrier of faulty copies of these genes, removing the tissues in which these cancers could arise. The approach was drastic, but one that was recommended by some cancer specialists. Just a few years later, women in similar circumstances have a wider array of choices, including screening for changes in their phenomic blood analytes and microbiome to monitor for early warning signs of cancer transitions.[8]
- At this time, we know of approximately twenty additional gene variants that present high risks for cancers. Individuals with these variants can be followed closely for cancer transitions. For example, Lynch

syndrome, which causes colon and endometrial cancers, arises from different variants in six DNA repair enzyme genes. These variants cause the DNA to mutate and confer about a 60 percent lifetime risk for both cancers, which usually arise between the ages of 45 and 60. Yet with regular colonoscopies and minor ablative surgery for malignant intestinal nodules, one can avoid colon cancer. Similar approaches can be used with endometrial cancers. These genes confer cancer risks of one in 2,000 to one in 10,000—millions of potentially avoidable cancers across the globe.

Although white individuals of European descent make up about 15 percent of the global population, they comprise about 80 percent of all participants in genome-wide association studies.[9] These studies are thus poorly generalizable across diverse populations, leaving many vital questions unanswered. Addressing this need has become a major rallying cry in the scientific community as researchers and major funding agencies push to make genomics research applicable across the broad swath of human diversity. What are the risks for genetic diseases in different racial groups? Many are still being worked out. Are different variant genes involved? Yes, and a lot more needs to be done. These are critical questions; adequate genome and disease correlation data must be generated for all human races. In Chapter 11, we will discuss one powerful new approach to the generation of personal data clouds that will reflect the racial diversity of the US population.

What if a gene like BRCA2 only increased the risk of developing a type of cancer in a specific ethnic group—for example, in South Asian women under the age of 40? Since cancer genes can manifest differently in different ethnic groups, differing blood analytes or phenomic measurements may reflect the onset of the disease. It will be important to understand how different genes manifest diseases in the context of distinct family histories. In other words, the disease could be the same, but early signals could be different. We can ask the same kinds of questions when it comes to how men and women differ in their susceptibilities to disease. As researchers around the world build and analyze more individual data clouds from genetically diverse groups, we will learn how genome and phenome variations correlate with disease in ways that better represent the human population.

Genome variants can increase the likelihood of disease and contribute to nutritional deficiencies, which can be corrected with specific diets and supplements. Knowledge of these variants can suggest ways to optimize an individual's exercise regimen for health or peak performance.[10] They can predict predisposition to injuries and point to ways of preventing them.[11] At least seven different genetic variants predispose individuals to common musculoskeletal injuries, and six inform specific types of athletic performances. For instance:

• A variant of the gene APOE known as G-219T has been associated with a threefold increase in the risk of concussion in athletes. Knowing your child is a carrier of this variant could inform your decision if they express an interest in sports with high concussion rates, such as football, soccer, or hockey. Tennis might seem a better option—unless, that is, your child has certain variants in two collagen-encoding genes, COL1A1 and COL5A1, which have all been linked to increased risk for tendinitis, a disorder common to swimmers and tennis players that results in pain, swelling, and impaired movement.[12]

• Two other genes—MMP3, which is involved in connective tissue wound repair, and TNC, which encodes for a protein essential to providing structural and biochemical support to cells—have also been linked to increased risk for tendonitis. But these variants, along with most others that make someone more likely to suffer from an injury, have actionable possibilities, generally involving exercise, that can allow one to mitigate their limitations. One study showed that plyometric jumps and tool-assisted self-massage may impact exercise recovery.[13] So tennis might be back on the table if you know what to do to prepare.

• A gene variant controlling high maximal oxygen uptake known as VO2 max could potentially lead us in the direction of a drug that could confer some of the benefits enjoyed by "superathletes," whose VO2 max rates are often much higher than those of the rest of us. A medicine like that could be extremely useful for treating patients with chronic pulmonary obstructive disease or heart failure—and it's reasonable to wonder what such a medication could have done to help people struggling to breathe after contracting COVID-19. Such a drug might even offer a temporary boost in VO2 for an older person who wants to go on a long hike.

• Hundreds of gene variants—including several that are clinically well validated—have been reported to modify a person's response to frequently used drugs, causing them to be metabolized too rapidly or too slowly, leading to ineffectiveness or long-term toxicities.[14] The resulting conditions include clopidogrel resistance, warfarin sensitivity, warfarin resistance, and malignant hyperthermia. The new field of *clinical pharmacogenomics* promises to teach us how our individual genetic variants and their associated drug toxicities can affect our choices for appropriate drug treatment. More than one hundred of these variants have already been identified. These genetic predispositions should be known to your physician so that they can choose appropriate drugs that are more likely to work for you.[15]

Understanding Genetic Risk

Genetics that affect the risk of disease or injury fall into three categories: dominant genes, recessive genes, and polygenic risk scores.

Dominant genes are responsible for diseases that can be caused by a single copy of the "bad" gene. A key feature of dominant genes is penetrance—the frequency with which they lead to disease. Some genes are highly penetrant, always leading to disease. This is the case with Huntington's disease, a neurodegenerative disease in which the one bad gene inevitably leads to neurodegeneration. Most people who are at risk of Huntington's know they are at risk because they had a parent or grandparent who had the disease. Unless they are genetically tested, there's no way to know whether they inherited the bad gene until symptoms begin to manifest, generally in their sixties or later. Choosing genetic testing means potentially living the rest of your life knowing that, at some point, you will begin to slide into a spectrum of cognitive, muscular, behavioral, psychological, and emotional symptoms. However, it also offers the opportunity to plan and prepare for what is to come.

Having witnessed the devastating effect of the disease on many of his family members, Huntington's disease researcher Jeff Carroll, who carries the Huntington's gene, and his wife opted to use in vitro fertilization so that their children would be unaffected. Fortunately, doctors can now identify whether embryos have the defective gene. Consider the implications of such a decision: a disease that has existed in Carroll's family for untold generations will no longer be a problem for his children or their children. He has eliminated the Huntington's gene from his family lineage with a single decision informed by modern science.[16]

Thankfully, most dominant genes have lower degrees of penetrance, meaning they elevate the risk, but disease is not a certainty. This is the case with BRCA1 and BRCA2, the breast and ovarian cancer genes.

In recent years, a tremendous amount of research has centered on how environmental factors can affect the expression of genes and the likelihood a genetic predisposition might trigger the expression of a disease. Scientists have found that air pollution has a particularly powerful effect on gene expression, potentially cranking up the expression of genes that make us more susceptible to diseases such as asthma or chronic obstructive pulmonary disease.[17] As a research community, we've just begun to explore how these environmental associations impact disease penetrance.

Another less frequently studied class of factors that might be more powerful are disease-modifier genes. These are present in the genome at a distance from the gene they modify. Everybody's inherited genome is different (not even monozygotic twins are completely genetically identical). Modifier genes might help explain why someone can inherit a dominant gene for disease from a parent but have a very different experience with that disease than that parent—and why, even if we could eliminate the expressive power of environmental factors, not even identical twins would experience an inherited disease in exactly the same way.

The traditional understanding of genetic diseases rests upon the fundamental notion, going back to geneticist and mathematician Gregor Mendel, that a single copy of a defective gene can cause autosomal dominant disease. An individual needs two copies of a defective gene, however, to inherit a so-called Mendelian recessive genetic disease like in sickle cell anemia, cystic fibrosis, Tay-Sachs disease, and hemochromatosis.

Nine people out of the 5,000 who joined Arivale were identified as having two bad copies of the hemochromatosis gene, known as HFE, but only three of them had high blood levels of iron consistent with the disease. Clearly some modifying genes, environmental factors, or both were suppressing the manifestation of the disease in the other six individuals who had two defective copies of the gene.

Here's where things get interesting: forty-five people in the group had inherited just one copy of the defective version of HFE. Genetics alone would suggest that they should not have been at risk for hemochromatosis, yet nine of these patients had levels of iron in their blood consistent with the disease. This could be a coincidence—there are lots of ways to end up with high levels of iron in one's blood, including liver disease and excessive iron in the diet—but statistical analysis of that possibility suggests something more profound, that modifier genes may promote a recessive disease in those with only one bad copy of the gene.

There are about 7,000 Mendelian recessive diseases, all of which supposedly require two bad gene copies to manifest themselves. Is it possible that just a single bad copy is enough to trigger the disease under special circumstances? And what could those circumstances be? Could environmental factors, stressors, or modifier genes be at work, triggering rather than suppressing the disease? This is one reason complete genome sequences of every individual are so important. We can examine our genomes for each of the 7,000 Mendelian genes to determine whether we have potential gene-damaging mutations, follow affected individuals closely throughout their lives to identify their earliest disease transition, and, if possible, reverse the transition.

The third category of genetic diseases occurs as the result of a polygenic combination of many—up to thousands—of variant genes, each contributing minutely to the total genetic risk. These variants are identified by carrying out genetic studies on the genomes of populations ranging in size from tens to hundreds of thousands. The comparative risk of complex genetic diseases such as these can be represented through what are known as polygenic risk scores, which sum up the risk for each individual of contracting any given disease. Understanding what these genetic risk differences mean is critical to extending our healthy lives. There are more than one hundred diseases currently associated with polygenic scores, including most of the common diseases.

The data clouds from Arivale turned out to be highly useful and instructive in helping us begin to understand the reflections of high polygenic scores in individuals. Early on, we realized that using polygenic scores as a filter to analyze the data clouds for each individual could be incredibly powerful. While most diseases are only acquired by a small fraction of people, we can calculate a polygenic risk score for any genome. Thus, every person's data can be relevant to the study of every disease because we can learn how high-risk individuals differ from low-risk individuals and how both differ from those of intermediate risk.

These genetic risk scores give us a lens through which to consider what differentiates those who have a higher risk of contracting any given disease from those who have a lower risk before they ever develop symptoms. That's important because biomedical research most often studies late-stage disease progression against a control group. Within that control group, the vast majority of people don't have the disease being studied and most won't ever get it. Prior to the existence of data clouds, control versus diseased patient juxtapositions were the best comparisons that could be made for clinical trials. Now we can stratify all patients by genetic risk to select more appropriate control patients, making clinical trials more focused and effective in revealing fundamental new disease insights.

The ability to evaluate genetic risks in one hundred or so diseases means we don't have to wait for decades to see disease arise in any given group of people and then go back to evaluate warning signs. We can start immediately with high–genetic risk patients and look at their changing phenotypic signs through the lens of an already-known genetic landscape. We can learn about disease onset before anyone ever gets clinically detectable transitions by tracking phenomic changes.

As noted above, scientists have developed risk scores for more than one hundred different diseases, each of which will continue to be refined as we develop our understanding of polygenic risk for these and many other diseases. As it is, tremendous benefits can be gleaned from knowing these scores. If you know you are at high risk for a certain disease because of family history, you may be able to use longitudinal deep phenotyping to identify the earliest instance of a wellness-to-disease transition, just as we can follow individuals with elevated genetic risk for disease. We can also see how blood analytes, such as LDL cholesterol or microbiomes from a stool sample, change over time in high-risk individuals.

How do we do that? By analyzing the ever-changing phenome to determine what molecular systems are different in those with high genetic risk for a particular disease. We can assess blood analytes, gut microbiomes, and digital measurements. From these measurements come deep insights into predisease states, assessing and slowing aging, and detecting the earliest wellness-to-disease transitions.

Reading Our Blood

A drop of blood is a universe unto itself, filled with proteins, metabolites, and cells. The changes in the levels of these molecules or cells can give us information about which biological networks are in flux, and these insights can be acted upon swiftly. As we noted in Chapter 2, we saw an example of this during the induction and transition from wellness to disease of prion-induced neurodegenerative disease in mice.

Doctors have been using blood analyte panels for decades. The comprehensive metabolic panel is a common test that measures fourteen different substances in a person's blood, including glucose, calcium, several proteins, and bilirubin, a waste product made in our livers. But there is nothing magical about any of those fourteen substances or, for that matter, those tested for in other common blood panels. Yes, any one measurement may indicate a nutritional or vitamin deficiency, or a surge in a certain liver enzyme that can be indicative of a particular problem, but it is the combination of blood analyte measurements that best serves to alert us to the presence of disease or predisease.

The most common predisease condition is prediabetes. Over a third of all American women and a bit more than 40 percent of all American men are prediabetic, which means that tens of millions of Americans are on this early path to diabetes, with potentially severe consequences. They are increasingly being joined by individuals in other parts of the world, especially the United Kingdom, Asia, and Latin America, as the diabetes-causative American diet is spreading.[18] Most individuals who are prediabetic do not realize it, despite the fact that assessment only takes a simple blood test. Anyone with a fasting blood sugar level between 100 and 125 milligrams per deciliter is considered prediabetic. Measures can be made to monitor for insulin resistance, elevated hemoglobin A1c, and other blood markers of diabetes as well.

For many years, we have known that childhood blood sugar levels can help predict who will wind up prediabetic (and often eventually diabetic). Children who are not prediabetic but usually on the high end of blood glucose levels are generally considered to be healthy. But these children are more than three times more likely to be diagnosed with prediabetes as adults. This is not the only indicator that we can use to predict a greater risk: blood sugar levels indicative of prediabetes are often preceded by increases in other substances many years and even decades in advance. Small increases in cholesterol and certain types of fat and calcium in the blood during youth have been associated with high blood sugar much later in life.[19]

Clinical chemistries can also provide insight into pre–cardiovascular disease, pre-obesity, and pre–metabolic syndrome. (As many as one in three Americans have metabolic syndrome, which is linked to obesity, a sedentary lifestyle, and insulin resistance.)[20] Emerging tests can detect proteins that are indicative of early cancerous growths, too.[21] In our work at Arivale, we discovered that levels of particular analytes indicated that disease transition had begun long before a clinical diagnosis in several types of cancer. A protein coded for by the gene CEACAM5, for instance, was a persistent longitudinal outlier as early as twenty-six and a half months before the diagnosis of several forms of cancer. CALCA, a biomarker for medullary thyroid cancer, was hypersecreted in individuals with metastatic pancreatic cancer at least sixteen and a half months before diagnosis. And the protein coded for by a gene called ERBB2, which is known to be associated with several autosomal recessive diseases, spiked in metastatic breast cancer patients up to ten months prediagnosis.[22] These findings convinced us that most diseases—perhaps even all of them—have "pre-warning signs," some simple, some more complex, that can be identified by quantification in blood.

When we tell people this, they are often flabbergasted. "Why aren't we making more of an effort to identify these warning signs?" a friend recently asked us. Our answer was likely unfulfilling: it is because of concerns over false positives, costs, and some antagonism toward the concept of prevention in general (the healthcare system makes its money on disease, not on wellness or prevention). At some level these concerns are valid, even if in our view they are quite easily negated by advancing technology and relatively simple math. (Higher numbers of measurements, taken all at once, can reduce false positives by evaluating potential diagnoses based on a number of coordinated signals at once—as we will explain.) What is not valid, in our opinion, is the fact that many of those who are resistant to these sorts of interventions have not kept up with the latest science.

Measuring Our "Real" Age

It's not just individual diseases that can be measured and tracked—and intervened in—through a complex analysis of the genome, phenome, and other measures of health. Biological aging—the deterioration of cells and systems resulting from genetic, epigenetic, lifestyle, and environmental factors—has correctly been described as the greatest risk factor for most chronic diseases. Biological age is the age your body says you are, as opposed to your chronological age. At one time, the best way to get an idea of how much aging might impact a person's susceptibility to transition and disease was to simply add up the years they had lived. Doctors have been turning to chronological age for as long as medicine has existed, but it is a very inexact metric.[23] A person who is 50 years old could have a substantially different biological age—older or younger—that may dictate a different course of action for the treatment or pretreatment of disease. We will discuss biological aging in more detail in Chapter 7, but the important takeaway for now is that biological aging is yet another data point that should be incorporated into any individual's personal data cloud for the purpose of predicting and preventing a transition to disease.

Before we can consider biological aging, however, we must be able to measure it. Multiple researchers and research teams have come up with ways to do so, and we have too. We developed a measurement for biological age using 1,200 blood analytes—proteins, metabolites, and clinical lab tests—across 3,558 individual data clouds. This analysis gave us a very deep view into biological age across many systems. We later arrived at a biological age estimator that, in a simplified form, is being used as an overarching metric for wellness by Thorne, a health company for which Nathan is now chief science officer.[24]

Measuring Wellness

To better understand how to interpret all of these signals, we can't build data clouds only for people who are sick. We must commit to deep phenotyping of individuals who are very healthy, creating a growing number of genetic and phenomic baselines from which to identify the important factors impacting wellness and the earliest signals of transition to disease. Such studies will have a profound effect. Billions of a relatively narrow range of blood tests are run each year impacting treatment decisions for people around the world. Assessments of these tests typically focus on measuring a particular lab value and then determining if it is "too high" or "too low" based on an "acceptable" normal clinical range. But these ranges, called "reference intervals," focus only on major deviations that might warrant intervention with drugs for a subset of the population.

Our data clearly show that the expected values for many blood analytes differ substantially from the norms as a function of personal genetics. This conclusion aligns with the growing recognition that widely used clinical ranges generally fail to reflect ethnic diversity (as we noted earlier, white subjects are highly overrepresented in clinical trials). When researchers from the University of Hawaii investigated the accuracy of these guidelines for treating disease in an ethnically diverse population, the dismaying—albeit unsurprising—finding was that racial and ethnic subpopulations have significantly different reference intervals.[25] There is clearly a huge need to refine our health benchmarks to represent this diversity. By coupling individual data clouds with new techniques in genomics analysis, we can deconvolute this barrier to individually appropriate care, allowing for increasingly accurate treatment for people of different racial backgrounds and ultimately making it possible to offer personalized care that aligns to each individual's unique genetic profile.

Cholesterol level is a good example of a measurement with incomplete and inexact "normal ranges" in need of reform. Most personal physicians can recite from memory the standard reference intervals for this widely measured lipid and will use the ranges of LDL and HDL cholesterol to decide whether to recommend statin drugs such as Lipitor to lower LDL, commonly known as "bad" cholesterol. Although Lipitor is one of the most prescribed drugs in the world, this is not a decision most physicians make lightly, for along with its benefits, the statin has a wide variety of toxicities including muscle wasting, diabetes, loss of appetite, confusion and memory problems, drowsiness, fever, and jaundice. But research conclusively shows that not everyone experiences the expected impacts of "high" and "low" levels of LDL and HDL in the same way.

By combining information from clinical chemistry and polygenic risk scores for LDL cholesterol, we can demonstrate a strong genetic effect that influences both LDL and HDL baselines. The integration of genetics and blood measures introduces a fundamental new concept to their interpretation—the difference between the measure in your blood and the genetic prediction of your expected baseline. Strikingly, we found that, when individuals had the same high LDL levels, those whose genomes predicted a lower genetic risk for LDL were much more likely to be able to lower their LDL with diet, exercise, and other simple measures than those whose genetics predicted a higher genetic risk for LDL cholesterol.[26] In fact, for people in the top 40 percent of predicted genetic baseline for LDL, we saw no significant reduction in LDL throughout their journeys in Arivale. Across the board, they needed statins to bring down their LDL cholesterol levels. In stark contrast, those in the bottom 40 percent were able to lower their LDL cholesterol levels markedly with just diet and exercise, meaning they didn't need drugs at all, even though a doctor reading their levels might reasonably conclude they were in need of drugs based on the clinical reference ranges.

The same was true for increasing HDL, the "good" cholesterol. To our knowledge, this was the first concrete demonstration that genetics could predict the outcome of a lifestyle intervention—a key point for building the case for data-driven, genetic-based wellness programs. Finally, we had concrete evidence that the combination of genetics and blood measures offers a much better predictive indicator as to what treatments and therapies will work best for each individual—the real measure of success for personalized medicine.

Unique Paths for Wellness

All of this suggests that, for many polygenic diseases, we can treat high-risk individuals very differently from low-risk individuals. This will enormously expand the number of possibilities for actionable interventions long before a disease begins to manifest. It also suggests a future in which many fewer people will be exposed to the potentially toxic side effects of medication designed for late-stage disease intervention that too often brings them no real benefit.

We have used data clouds to map out the manifestation of genetic risk for fifty-four different diseases and conditions, from breast cancer and chronic kidney disease to asthma and Alzheimer's.[27] Every one of these fifty-four diseases and conditions had at least a few blood analytes, out of the 1,200 we tested every six months, that increased or decreased with escalating genetic risk. Collectively, this amounted to approximately 800 blood molecules whose concentrations changed with increasing genetic risk in these fifty-four conditions. We sampled many different biological systems in the context of these genetic risks—each of which has the potential to provide fundamental new insights about disease. With increasingly comprehensive data clouds for increasingly larger numbers of people, the extent of these associations will deepen. What makes finding these associations important is that each offers a clue as to what systems these disease genes are affecting and provides candidates that could be monitored for early detection of transitions or identified as drug targets.

One example is the associations identified between genetic risk, blood analyte concentrations, and amyotrophic lateral sclerosis, or ALS, also known as Lou Gehrig's disease, a neurodegenerative disease that is generally fatal within one to two years of diagnosis and for which there is no cure. This disease causes paralysis in certain muscles, including those that control swallowing, while leaving one's mental abilities intact. Tragically, those afflicted by ALS succumb to increasing physical paralysis while remaining entirely mentally lucid.

Using data cloud maps, we found that ALS was negatively associated with unbound ethylenediaminetetraacetic acid, or EDTA, which is commonly used as a blood preservative. Because it is a blood additive, our first instinct was that this could be a spurious correlation. Maybe the cloud had pointed us in the wrong direction. That can happen sometimes, which is why some physicians express concern that overtesting could lead to false positives. In this case, the additional data came in part from a lack of association between EDTA and any of the other fifty-three genetic-risk conditions tested. The association existed uniquely for ALS.

When we dug deeper, we learned that one of the known properties of EDTA is that it binds to heavy metals such as lead and mercury—and exposure to heavy metals is a known risk factor for ALS. Remember, we are studying the manifestation of genetic risk for the disease, not comparing individuals who actually have the disease. So, while heavy metal exposure is a risk factor, there was no expectation that those at higher genetic risk for ALS would necessarily have higher heavy metal exposures. What this finding suggests is that people at higher risk of ALS may be less able to clear these heavy metals from their blood, leaving them present in higher concentrations. And what that tells us is that if you have a genetic risk of ALS, you are likely to be more sensitive to environmental heavy metals. It further indicates that anyone with a family history of ALS or associated risk factors should work hard to avoid heavy metal exposures and think twice before consuming swordfish or tuna, both of which have high levels of mercury.[28]

Little by little, it is becoming clear that every person on this planet has a unique path for maintaining wellness, identifying and reversing wellness-to-disease transitions, and fighting disease. As we come to understand the health effects of individual genetic dispositions, we will almost certainly be horrified that we ever let very old, small, and homogenous samples of test subjects determine how we should fight diseases for everyone.

Unlocking the Secrets of the Gut

What is equally baffling is why we have so long ignored what is happening in our gut. The human body is a complex ecosystem with roughly the same number of microbial cells as human cells. Distinct microbiomes are associated with the gut, skin, and oral cavity, and we have about 100 times more microbial genes present in our bodies than we have human genes. All microbiome species have the capacity to impact the biological systems that surround them. Just as genes and blood analytes interact and impact our health, so do all these microorganisms existing on and within us—in ways we are only beginning to understand.

There is one microbiome we have a bit more of a handle on than the others: the one in our gut. Over the past decade, researchers have shown that our gut flora impacts everything from our metabolism and nutrient absorption to healthy immune development, effective drug usage, and even mental health.[29] These impacts are primarily the result of bacterial digestion and waste. By way of example, *Bacteroides thetaiotaomicron* can convert sulfur-containing glucosinolates—which are responsible for the pungent smells of plants like broccoli, cabbage, and horseradish—into isothiocyanates, which are known to have protective effects against some forms of cancer.[30] One experiment showed that an individual's gut microbiome was able to transform the chemical structures of fifty-seven drugs, 13 percent of those tested.[31] Intriguingly, which drugs were altered in this way differed from person to person. Thus, a drug that you are taking might not even be reaching your relevant biological system, instead it is being metabolized away by your microbiome, inhibiting effectiveness or generating deleterious side effects. Each of us can benefit from these emerging insights. In the past few years, we've learned more about human microbiomes than we've known in the previous three and a half centuries. Despite the immense progress made in this area, however, there is still no clear consensus of what constitutes a "healthy" gut microbiome.

Here's what we do know: each person's gut flora is unique, varies in space and time, and becomes more individualized over time. This became apparent in the people we were working with at Arivale. Even among the youngest participants, and among those closely related or living together (in which case the exchange of microbiota is common), gut diversity is unique from person to person. But around the age of 40 or 50, that uniqueness—the distance to the nearest microbiome from another person in the cohort—starts to increase.[32] That is to say, your microbiome becomes less similar to anyone else's.

Interestingly, this increasing uniqueness appears to be a function not only of living longer but also of living healthier. We do not see the rise in uniqueness in individuals who contract serious diseases, take multiple broad-spectrum antibiotic medications, or self-report lower health across a number of metrics. Those who stay healthy have microbiomes that continue to evolve, becoming more personalized. These findings were validated beyond Arivale in two other cohorts, representing about 10,000 individuals in all. When we took this uniqueness score and validated it in another cohort of men between the ages of 80 and 100, we found that the uniqueness of each individual's microbiome actually predicted who was at the highest to lowest risk of death in the next four years. What this means and how it impacts lifelong health is still being worked out, but we were fortunate to receive a Healthy Longevity Catalyst Award from the US National Academy of Medicine in 2020 to further study this increasing microbiome uniqueness as a function of aging.

What we can say about gut health now is that there is no "right combination" of microbes for everyone. Two people could carry very different microbial communities but have similar health or disease outcomes that are nonetheless associated with their microbiomes. This makes it hard to extrapolate clear connections between our microbiomes and our health, and makes it even more challenging to create interventions that work for many people. A study published in *Nature Medicine* used machine learning to compare microbiomes across multiple provinces of China and discovered several patterns that were predictive of health.[33] But when the researchers came upon a pattern for health in the microbiomes of people from one province and then applied it to those in a different province, it failed. The microbes were too different from place to place. The health predictions were place specific.

While the species of microbes that inhabit our guts vary widely from person to person and from region to region, the metabolic niches that any microbiome has to fill to contribute to human health is largely consistent. In the case of disease resistance, for example, a healthy microbiome can help "train" the immune system to fight off infections and protect it either by consuming substances an infectious pathogen would need to spread across the body or by creating chemicals needed to ward off disease.

Just as there are many different ways of teaching children to read, write, learn math, or tie their shoes and no student's path is quite the same, different assemblages of bacteria can come together to train the immune system in this "immune system–microbiota alliance." And just as a trail of documentation in the form of report cards attests to a student's readiness to transition from one grade to another, a trail of molecules—the chemicals that have been excreted from microbes doing exactly what they are supposed to be doing to ward off disease—in the blood may affirm that our microbiota are doing their work. We may be able to collect a lot of information about the microbiome using a blood test that allows us to see what the microbiome is contributing in the form of blood metabolites.

With this in mind, Nathan and Lee and some of our ISB colleagues set out to see if we could predict whether someone has an overall healthy diversity of microbes in their gut without ever actually looking at those microbes. We hypothesized that by simply taking stock of the analytes in a simple blood draw, we would be able to accurately estimate whether any given individual had a diverse (and thus generally healthy) microbiome or a less diverse (and thus generally less healthy) one. Using the Arivale data sets, we tested different kinds of measurements, including common clinical panels, larger arrays of proteins, and metabolites, to determine whether they could be used to accurately predict the diversity of the gut microbiome.

Applying machine learning to each of the data sets, we found that the strongest predictors by far were the metabolites—those small molecules in the bloodstream that serve as fuel, help convey signals from system to system, and stimulate or inhibit enzymes. This makes sense given the important role the gut microbiome plays in digestion, metabolism, and nutrient absorption. We are now able to predict degree of gut microbiome diversity—a measure of the species and how evenly they are distributed—using just forty blood metabolites, as we showed in *Nature Biotechnology* in 2019.[34] This model has allowed us to generate robust predictions of gut microbiome diversity across different disease states and has shown consistent results in a separate population of individuals.

There is still a lot we don't know about the gut microbiome. We've only scratched the surface when it comes to understanding the effects different species of gut bacteria have on the bloodstream. We don't yet know nearly enough about the impacts that different microbes have on one another or how those impacts play out when it comes to wellness and disease. What we do know is that the gut microbiome plays a vital role in human health and should thus play an important role in healthcare. For that to happen in the near future, we need to start measuring each person's microbiome and acknowledging its essential role in their health data cloud.

Separating Signal and Noise

Right now, wherever you are and whatever you are doing, if you are carrying a phone or wearing a smartwatch, or if you are within range of a smart speaker or in view of a camera, you are actively creating personal data. Your dynamically changing biological systems are also creating data, reflected in changing concentrations of blood analytes and microbiome species. Much of these data are noise. A lot of it is not at all helpful for understanding your health and well-being, at least not right now, but much of it could be valuable down the road. Yet we are allowing nearly all of it to disappear into the digital ether—in any event, it is not readily accessible to us as we seek to optimize our health journey.

So how do we separate the signal from the noise? Let's start with what is obviously helpful. Smartwatches and smartphone apps can now record a person's heart rate, stress levels, sleep quality, blood pressure, heart rate variability, altitude gained, and daily steps, among a growing list of other measures of digital health.[35] The accuracy of these measurements is improving so quickly that top-of-the-line health monitoring devices and apps are often eclipsed by newer, better products in just a matter of months. Using the same monitoring technologies, "digital phenotyping" can detect patterns from text messages, typing, movement, posture, and speech and use these as diagnostic tools, providing early warning for psychological distress, neurological problems, or signs of an impending stroke, heart attack, or disease. This is all potentially helpful data.

We are entering a brave new world of biomedical monitoring. Yet despite the fact that we can easily track people's activity level, sleep, pulse, and blood pressure—almost to the step and the minute—the way most doctors access this information is by asking patients to self-report. A lot of doctors don't even do that, limiting their information gathering to asking questions such as "How often are you exercising?" and "How are you sleeping?" Research shows that these questions are all but worthless because people aren't very good at estimating how much sleep they get and are even worse at accurately reporting how much they are exercising.[36] Yet, along with genetic considerations and diet, these are the two most important factors for determining what treatments and therapies will work. A doctor prescribing the drug ranolazine to treat a patient's chest pain, for example, might very reasonably be interested in how much a patient is exercising, since elevated body temperature and elevated heart rate—two physiological hallmarks of exercise—might decrease that drug's efficacy.[37] A doctor prescribing corticosteroids for inflammation might likewise first want to know how their patient is sleeping, since these drugs can cause insomnia—potentially creating an even greater health concern than the inflammation—and the doctor might want to consider alternative treatments that don't impact sleep.

Physicians don't have to take a patient's word for it when it comes to heart rate and blood pressure. Those are the very first things measured when a patient walks into the examination room. But these data points offer no more than snapshots of a single moment in a patient's life. A person's resting heart rate in the doctor's office tells us nothing about how their heart responds to activities like racing through the airport to catch a connecting flight, a tense interaction at work, or a heated exchange with their spouse. Blood pressure taken at a doctor's office may be a poor reflection of our normal blood pressure or what happens in our arteries when we are cut off by a dangerous driver, ignored by our kids, or insulted by a boss.

Many people who have been diagnosed with atrial fibrillation are all too familiar with the frustration of racing to the doctor's office when their heart is out of control, only to have it settle to a normal pattern in the waiting room. This is common. Yet many doctors are hesitant to treat a patient for paroxysmal atrial fibrillation (A-fib), an arrhythmia that comes and goes, until they actually see the problem themselves. This is tremendously challenging because early interventions can be lifesaving for people with A-fib. Today, electrocardiogram-enabled smartwatches are changing the game, giving patients the tools they need to show their doctors what is happening with their hearts outside of the examination room.[38]

Measurements like these are invaluable for doctors and patients alike, and not just when it comes to deciding an initial course of treatment. Recall that the Arivale data demonstrated that genetics and blood analytes could provide a much better understanding of what sort of interventions would work best for people who would likely just be prescribed Lipitor under the regime of standard reference intervals. Now that we know that some people respond better to drugs and others respond better to lifestyle changes—and now that we are getting much better at figuring out who is who, thanks to genomic and phenomic analysis—we can prescribe interventions accordingly. This practice alone stands to improve the lives of many patients who would previously have been taking drugs with plenty of negative side effects but very little, or even no, positive impact. But a doctor who considers a patient's data cloud to make a treatment recommendation would be wise to keep watch over digital health measures as well, because even if lifestyle changes would be better, a person who doesn't consistently follow through with exercise as directed may turn out to be better off taking a statin after all.

This is why data clouds are so important. They allow us to analyze and integrate diverse pieces of information to gain insights we could not otherwise have reached and to individualize personal care in a way that was simply impossible as little as ten years ago. This is the essence of why data-driven health is so revolutionary.

The Next Steps in Data Collection

We are still at the very early stages of understanding what data clouds will reveal about our health and how they will enable us to predict and prevent disease. Discoveries are emerging rapidly from around the world, and efforts are scaling up. China is taking a leadership role, having formed a Human Phenome Consortium and committed hundreds of millions of dollars to deep phenotyping, enabling more and more precise measurements of the human body, creating ways to finely deconstruct various physiological systems, and identifying precise and person-specific wellness and disease interventions. It is not unusual for those who believe in this vision to compare the consortium's scope and potential to other world-changing advances.

"With ship and navigation technology, humans can explore the world; with astronomical telescopes, humans can explore the universe," Li Jin, a member of the Chinese Academy of Sciences and a co-director of the consortium, explained at the second International Symposium of Human Phenomics that we attended in Shanghai in 2018. "With human phenome research, we can explore the inner world of the human body."[39]

Closer to home, we are in the process of formulating and implementing a million-person genomics and longitudinal phenomics effort through the nonprofit Phenome Health, led by Lee. The goal of this project is to return thousands of actionable results for patients in a way that acknowledges and honors the importance of delivering insights to all groups across the population. (The program will be discussed in Chapter 11.) The effort complements the NIH's All of Us Research Program, which is focused on carrying out genomic sequencing for one million patients and making samples available for research programs to build on this foundation.

These efforts will transform our understanding of health and disease, but each one can only be as good as the accuracy and range of the information that feeds it. So how do we get that part right?

4

Measuring and Tracking Health

Why Deep Monitoring Is Essential to the Healthcare Revolution—And How It Might Backfire

Warren Buffett wants to buy you the car of your dreams. "You can pick out any car that you want," the billionaire investor writes in *Getting There: A Book of Mentors*. Go ahead and picture the car you want. The next time you return home, it will be waiting outside with a big red bow on it. But that's when you learn there's a catch: this is the *only* car you can ever have for the rest of your life. With that in mind, Buffett asks, "how are you going to treat that car?" You'll probably read the owner's manual before you drive it and stay up to date on oil changes and other maintenance. You'll go all out to make sure it lasts.

And yet, Buffett laments, many people don't show the same concern for their bodies. "You have only one mind and one body for the rest of your life," Buffett writes. "If you aren't taking care of them when you're young, it's like leaving that car out in hailstorms and letting rust eat away at it."[1] In Buffett's parable, you only learn the catch *after* you choose the car. But what would happen if you knew it ahead of time? Would it impact the car you choose or the features you'd want?

In recent years, American consumers have been holding onto their cars longer. According to a federal Department of Transportation survey, the average vehicle in 2019 was twelve years old, an increase of 4 percent in just a half-decade.[2] One big reason for this: today's vehicles are equipped with hundreds—sometimes thousands—of sensors. These include monitoring devices that let you know if the engine is running hot, if tire pressure is low, or how much fuel remains, down to the mile, based on current driving conditions. Newer models include cameras and radar that keep track of surrounding vehicles anytime the car is in motion.

These sensors appeared so gradually that many of us don't realize how much the experience of vehicle maintenance has changed—or how much we now depend on them. Modern engines are so complicated that mechanics generally have to plug the car's sensor system into a diagnostic computer to know what is wrong with them. On the flip side, as long as we're paying attention, we will be warned well ahead of time of any failure on the road.

If you treat it well and avoid major collisions, a car you purchase today could be with you for decades. That's still just a fraction of the time you'll spend with your body, which is yours from birth till death. Yet, for reasons no one can adequately explain, we haven't come around to demanding the same sort of care for our bodies as we do for our cars. In fairness, it is a much more daunting task to keep our bodies well than it is to keep our cars functional. But isn't it time we at least tried?

Preventive medicine has been getting more and more lip service in recent years, but it is still not practiced widely beyond "checkups," generic advice to eat well and exercise, and critical population health measures such as vaccines. This is starting to change, though. A lot of people now use their smartwatches to count their daily steps and monitor their heart rate or sleep. A flood of digital devices like the Apple Watch, Oura Ring, and Fitbit has made personal health measurements far more accessible. But we are still just scratching the surface of what we could know about our bodies at a molecular level.

Opportunities for Intervention

Let's briefly return to the simple model of health and disease we discussed in the Introduction (see Figure 4.1).

For the purpose of simplicity, let's assume most people begin in a state of wellness. At some moment, a process begins that could result in disease, often as the result of an external perturbation brought on by lifestyle choices, hormonal fluctuations, or environmental exposures, sometimes triggering a genetic predisposition. This is disease initiation, or a wellness-to-disease transition. Following its initiation, we can see a cascading series of changes—triggered by the emergence of disease-perturbed networks. This early state of disease progression, or

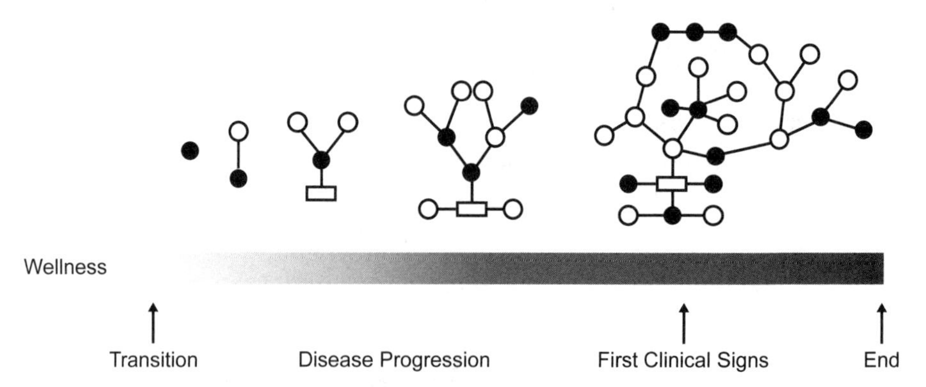

Figure 4.1. The complexity of disease-perturbed networks grows exponentially as the disease progresses, making it easier to reverse through early intervention.

predisease, ultimately leads to clinical signs and obvious (diagnosable) disease.

Let's consider some real-world examples.

Type 2 diabetes is a chronic disease that affects hundreds of millions of people across the globe. With the exception of those with type 1 diabetes, whose pancreases make little to no insulin (a very different challenge), most people's bodies control the glucose levels in their blood through the synthesis of insulin in their pancreas. Insulin controls the level of glucose in the blood. This is the state of wellness.

Long before type 2 diabetes is clinically diagnosable, some people's systems will begin to lose the ability to effectively control glucose levels. The initiation has begun, and now these individuals are in a state of transition. Because it is so common and has such obvious blood signatures, diabetes is one of the few diseases for which doctors can identify a predisease state. At this point, patients begin to exhibit insulin insensitivity, meaning that rising insulin levels in their blood are less capable of bringing their blood sugar levels down.

Prediabetes is easily identified through clinical measurements such as insulin resistance, fasting glucose levels, and glycated hemoglobin (HbA1c, often abbreviated to A1c). Unfortunately, we act on these measurements far less frequently than we should. Early interventions are far easier and far more effective than the more complex and generally ineffective therapies available to treat advanced diabetes. In most cases, all it takes to reverse prediabetes are some straightforward lifestyle choices, including a decrease in dietary sugar and an increase in exercise. These changes require some discipline but are generally simple and even pleasurable. Lots of people who choose to walk to work and cook at home for health reasons grow so fond of these choices that they would continue even if it weren't necessary for their health. And while some might not enjoy eating healthier foods and exercising more, it's hard to make a case that these are worse fates than a lifetime of diabetes.

Absent an intervention, the progress toward fully developed diabetes will continue until the body's inability to control blood glucose levels becomes so pervasive that clinical signs appear. In late stages, severe symptoms include a loss of pain sensitivity in the extremities, often resulting in festering cuts and sores. In the United States, nearly 15 percent of all people with diabetes will develop a foot ulcer at some point in their lives, and more than 20 percent of that population will require amputation.[3] While this risk is more pronounced for type 1 diabetes, it is significant for type 2 as well. A second complication of diabetes is gradual kidney failure, so that one has to resort to dialysis to remove toxins from the blood that would typically be removed by the kidneys. A need for dialysis is generally irreversible, so prevention is crucial.

Kidney failure, a common consequence of chronic diabetes, impacts more than 400 million people across the world and 37 million in the United States.[4] The human tragedy and economic costs of diabetes are staggering—some $327 billion is spent annually on its treatment in the United States alone. It's sobering to consider that this disease is almost entirely preventable and that we have actively, albeit not intentionally, built a society that ensures its prevalence by producing and heavily marketing unhealthy food and promoting sedentary lifestyles. By some estimates, only 10 percent of American adults who have prediabetes actually know they have it.[5]

This isn't a problem unique to diabetes. If anything, the fact that doctors recognize prediabetes as a diagnosable condition (it even has its own code in the International Statistical Classification of Diseases and Related Health Problems medical classification list, which is used by healthcare systems around the world) means we're further along when it comes to its preventative treatment than we are for most other diseases. That's good, because by the point at which a disease is clinically apparent, it has generally progressed to a stage of such complexity that treatments often fail. But why can't we have these sorts of warning signs for other diseases?

Well, we can.

Autoimmune disorders affect 50 million Americans. In some cases, as with most seasonal allergies, these health challenges are annoying but not life-threatening. But some autoimmune diseases, such as rheumatoid arthritis, celiac disease, or lupus, are extremely debilitating. Others, for instance, autoimmune myocarditis, which can cause dysfunction for which the only current treatment is a heart transplant, can be life-threatening.[6]

For reasons not completely understood, autoimmune disorders are on the rise. But even though treatments are improving, the way we diagnose these disorders hasn't changed much in decades. We wait for symptoms to appear and then jump into action. And that means we're missing out on a lot of opportunities to intervene early, when we can actually have a major impact.

An Arivale client named Lynn is a good example of the life-changing potential for early interventions for individuals with autoimmune disorders. Lynn had long maintained a healthy weight, exercised regularly, and eaten what just about anyone would consider to be a healthy diet: whole grains, lots of veggies and fruit, lean meats, and low fats. Not surprisingly, most of the biomarkers Arivale collected for Lynn indicated she was healthy. There were a few outlying negative indicators, however, including cholesterol levels that seemed a bit high given Lynn's healthy lifestyle, and interventions intended to lower her numbers didn't succeed. By itself, stubbornly high cholesterol could mean all sorts of things, but there was one more clue: an elevated level of thyroid-stimulating hormone.

"I followed up with my primary care physician, who was really receptive to the fact that I had been tracking my own health metrics," Lynn said. "She listened to my concerns and agreed that a further investigation was a good idea."

It's a good thing she did, because it turns out that Lynn had early-stage Hashimoto's thyroiditis, a condition in which her immune system was treating her thyroid as if it were an invading entity in her body and destroying its functional cells. It's important to note that Lynn wasn't experiencing symptoms that would lead most people to visit a doctor.

Like a lot of people entering middle age, she'd gained a little weight and had some feelings of tiredness, but none of that would have likely to prompt a doctor to investigate the possibility that she was dealing with Hashimoto's, which often isn't caught until it results in life-altering symptoms. Because she knew early, however, she could do something about it.

Here is where Lynn's story goes from being a celebration of the power of proactivity to a reflection of the currently frustrating state of healthcare. After making an appointment with an endocrinologist, Lynn set to work putting together her historical lab reports, organized into graphs, as well as writing down some specific questions she hoped he could address.

"After giving me thirty seconds to discuss my labs, without allowing me to ever get into my actual concerns or questions, he interrupted me and proceeded to give me a five-minute lecture on how labs are only useful for confirming disease diagnoses and for adjusting medication," Lynn recalled, noting that the endocrinologist only seemed interested in diagnosing something he could write a prescription for. He admonished her to return after she had developed debilitating symptoms that would be sufficient to merit a drug prescription, with its associated side effects. "He talked down to me as if I was a 10-year-old who didn't realize how the adult world worked. He kept repeating that 'there are people out there who like to run lots of tests just to make money off you.' . . . I found it very ironic that the same man who had wasted my time repeatedly lecturing me that others had made money off me by running 'unnecessary tests' himself made hundreds of dollars off my visit with him in which he provided zero helpful evaluation, information, or feedback." The physician's total dismissal of Lynn was all the more striking as Lynn had a PhD in a related biological field and was professionally well aware of issues surrounding the analysis of data—a fact the doctor was unwilling to hear.

Lynn was discouraged but not dissuaded. She read everything she could on Hashimoto's, including books that advised unconventional approaches such as diets aimed at the so-called "leaky gut syndrome," a controversial proposition that suggests that an unhealthy gut lining may become porous, allowing food, toxins, and bacteria to penetrate surrounding tissues, triggering inflammation throughout the body. The hypothesis that this is a common cause for autoimmune disease is unproven, but the most common solution—a nutritious, unprocessed diet full of foods known to reduce inflammation—is almost never a bad idea. "I decided to give it a try," Lynn said, "especially since conventional medicine so far only offered me the 'come back when your hair is falling out and then we will give you a drug' approach."

Lynn cut out gluten, dairy, soy, and sugar. She added supplements suggested by her primary care physician, including R-lipoic acid and curcumin phytosome, and she increased her consumption of healthy fats like avocado and flax. A few months into this new dietary regimen, her bloodwork showed a tremendous reduction of the antibodies that indicate Hashimoto's, as well as a normalization of the cholesterol levels that had first signaled that something might be wrong. "Not only had my thyroid markers improved dramatically, but my whole body health had improved," Lynn said. "By taking a more whole-person, proactive approach instead of just waiting to treat a fully dysfunctional thyroid, I improved not only my thyroid function but also many other aspects of my health."

Can every autoimmune disease be caught and treated early in such ways? Not yet. But scientific wellness provides hope that eventually we will be able to do so in many more cases.

Making the Invisible Visible

It's not exactly revolutionary to note that, in nearly every case of disease, the sooner we know that something is wrong, the less dangerous and more straightforward the interventions tend to be. The difference in the way we view twenty-first-century medicine really comes down to what "sooner" means. Rather than waiting for the earliest signs or symptoms, we believe that scientific wellness and P4 medicine should be defined by a constant search to prevent or reverse transitions as close as possible to their point of initiation. At that point, intervention is even easier, and transitions can be turned around before disease truly sets in and begins perturbing additional networks.

None of this is possible, however, unless we build the systems necessary to generate personal, dense, dynamic data clouds, as described in Chapter 3, and follow them longitudinally for each individual. Once those systems are established, it's in each patient's interest to use them and become an active participant in this process.

How do we do that?

We start by making the invisible visible. A number of processes that are fundamental to our health are invisible to us because conventional medicine believes that people can't or don't want to understand their own health or won't make the lifestyle changes necessary to better the odds of disease reversal. But just as you don't need to understand the principles of gasoline combustion to read a fuel gauge and take care of the needs of your car, you don't have to understand the cellular processes that lead to diabetes to read a glucose blood level measurement and appreciate what it means for your personal health. (In fact, not even doctors agree on what really causes insulin sensitivity at the cellular level; there is an ongoing debate about whether a person's body ceases to be as sensitive to insulin or whether the cells are so full of sugar already that the insulin signal can't push any more into them.)

By sharing more information with patients, we can move these important metrics into the realm of conscious awareness. That alone would represent a massive shift in how we care for ourselves and think about our health and well-being.

Next, we need to move the future into the present. It is well known in behavioral sciences that humans, when making decisions, are inclined to massively prioritize present pleasures over future benefits. For most of us, the pleasure of eating ice cream, Doritos, or a chocolate cake today outweighs the increase in disease risk that may manifest at some point in the future. We tell ourselves that we can eat an extra-large piece of pie today and we will make up for it with exercise tomorrow, but we often fail to make good on those promises—and now that we have a better understanding of the role of the gut microbiome, we know that isn't really how things work anyway. One of the great secrets to success in life is to invert this impulse and do the "good thing" today while telling ourselves we can do the "bad thing" tomorrow.

Personal data can move our awareness and understanding of the impact of what we are doing to ourselves into the present, which can be a powerful motivating tool. We know, for example, that people who weigh themselves regularly are more likely to lose weight.[7] Likewise, research has consistently shown that people who wear step-tracking devices take more steps.[8] Of course, not everyone chooses to make behavior changes that are beneficial to their health. We all do this to different degrees for a variety of reasons. Collectively, though, we are going through a transformation in our understanding of health. By engaging in these changes, you can play a role in remaking healthcare and our societal environment to support better choices, including the movement toward healthier food, standing desks in offices, making time for movement during the day, and helping to craft and support a healthcare system that effectively intervenes earlier in disease processes. We are in this together.

Finally, we have to use the dense data clouds we and others are collecting to develop personalized insights. It is one thing to know generally that something is good or bad for some percentage of the population. It's quite another to be able to see changes happening with immediate and personalized feedback, dynamically and unambiguously, in your own body as a direct result of your own choices—and to recognize that very different choices are often required for your friends or family. If billions of people are going to be asked to shift their expectations of what healthcare should be, they have to see improvements in their own lives—right away.

A technology that achieves all three of these criteria, which has recently become broadly available, is continuous glucose monitoring, or CGM. The technology is simple and painless: You simply wear a small CGM device on your arm for up to two weeks at a time. During this time, you can review your real-time glucose levels on your mobile phone. Although it was developed as a much-needed tool for people with diabetes who need to closely monitor glucose and inject insulin, CGM is a valuable and increasingly affordable tool that provides immediate feedback about your body's response and glucose levels after different meals (and snacks). This feedback, researchers have shown, can help people adhere to lifestyle modifications aimed at preventing a need for diabetes medications.[9]

While CGM is now in widespread use for people with type 1 diabetes, its utilization is still in its infancy when it comes to wellness. The scientific community has started trying to use CGM to answer questions about human metabolism that we wouldn't even have thought to ask previously. In one particularly innovative and influential study, Israeli researchers used CGM to take measurements of 800 study subjects every five minutes for a week—that's nearly 300 measurements per day on each individual, creating an incredible trove of data that simply wouldn't be possible with more traditional glucose testing technologies.

They discovered that different people's bodies respond in vastly different ways to the exact same meal. Some people's blood sugar rose more after eating sushi or a banana than ice cream. This variability arises in response to a person's genes, their lifestyle, and the composition of their microbiome.[10] Because of these tremendous differences, a diet aimed at stabilizing blood sugar levels that might work for some people may be ineffective or even detrimental in other cases.

The study showed that it was possible to create diets customized to different people's genes, levels of physical activity, and microbiomes while also accounting for personal taste or cultural dietary habits—a process that the study's senior authors, Eran Segal and Eran Elinav, explained in their book *The Personalized Diet.*

Consider how powerful this information could be. If you knew that, for whatever reason, your body was particularly sensitive to bananas, would that impact your decisions at the grocery store? If you could review your blood sugar levels with a quick glance at your wristwatch, would you do so before ordering dessert? CGM is just one of many technologies that can offer immediate personal biofeedback, visible in the present, to help us make decisions that are right for our own bodies. Integrating these data into longitudinal data clouds would give each of us powerful insights into our body's thousands of unique responses to food. This is something that could benefit everyone.

While we don't yet have long-term clinical studies showing that negative diabetes outcomes have been reduced with this approach, clinical trials have shown the effects on post-meal glucose response, sparking the development of companies like DayTwo and Zoe, which aim to offer consumers an in-depth view of dietary recommendations informed by their personal microbiomes. For most people, however, access to CGM and an increasing number of similar technologies that measure other physiological functions is a "pay out of pocket" privilege. Commonly used CGM devices include the Abbott FreeStyle Libre and Dexcom; personalized health programs that combine these devices with apps and health coaching include Levels and January AI.

Why should knowledge of this type be limited to those who can afford it? Why would healthcare providers not make it available to all their patients? Who wouldn't choose healthcare that ensures they are far less likely to get a disease over one that deals with the disease—at great expense—only after it has begun to destroy their life?

The disconnect between a commonsense approach to wellness and the limitations of current healthcare is why we believe something akin to a wellness revolution is on the horizon. But to access it, we will need to begin the transition to a world in which data clouds—reflecting our genes, phenomes, daily activities, ever-changing oral and gut microbiomes, and environmental exposures—inform personal health data for scientific wellness.

Rational Cause for Concern

At this point, let's pause to take stock of the objections to this data-driven path to scientific wellness. Physicians, scientists, ethicists, citizens, and politicians have raised valid concerns about cost, data privacy, lack of racial inclusion, government control of surveillance technologies, corporate intrusion into personal health data, genetic discrimination, and complexities of analyzing big data. There are certainly reasonable questions as to when and at what scale data-driven health is feasible and cost-effective. It would be foolish to suggest that there will not be unintended consequences—some of which we will one day say we should have seen coming. And it is reasonable to wonder whether the artificial intelligence strategies that will be called upon to bring this data-driven dream to fruition will always promote our best interests.

The only thing worse than ignoring these critiques would be to dismiss them out of hand. So, while there does not seem to be much argument against the basic goals of a data-driven, wellness-centric future for medicine, there are vigorous debates about the implementation of this vision. How we answer these questions will set a framework for the future of health.

As public health officials tried to get a handle on the COVID-19 pandemic in the early months of 2020, governments around the world weighed the benefits of tracking citizens' movements. Israel's parliament quickly approved, and the government rapidly deployed, a tool that allowed the country's security apparatus to track the phones of known COVID-19 patients, suspected carriers, and anyone with whom infected or exposed individuals had been in contact.[11] Meanwhile, in Russia, authorities deployed Moscow's 170,000 surveillance cameras—and massive facial recognition program—to catch people who were violating quarantine rules.[12] In France, once vaccines were rolled out, people were issued a "health pass," which they had to show before they could enter restaurants, museums, shops, or other public spaces.[13]

In the United States, the Trump administration marshaled tech industry leaders to strategize about how cell phone location data could be used to track the fast-moving disease.[14] Several companies used location data from the phones of college students visiting Florida for spring break, demonstrating in frightening detail how the students had exacerbated the spread of the disease across the nation by ignoring social distancing and not wearing masks.

The critical question that worries some is whether governments would use disease-related surveillance technologies to gain greater control in non-pandemic times. This has been a particular concern in China, which mustered unprecedented surveillance measures to contain the disease but also has been accused of immediately using those technologies to further limit fundamental civil and human rights.[15]

It doesn't take a global pandemic to get governments and companies interested in seeing how far they can reach into people's lives before there is pushback. Simple fitness trackers, of the type many of us wear to record our walks and runs, are creating data that can be used to determine your life patterns. And although companies generally promise that these data are "de-identified" before being shared, a person's patterns can often be used to recognize them. Do you wear a fitness tracker, or have you downloaded an app that requires the sharing of your location information? Have you recently posted a photo of yourself on vacation on any social media site? Does your current job have your name on its website? With those pieces of information alone, your identity can easily be drawn from data that have been disaggregated in an attempt to protect user privacy.[16] And when that's determined, every place you've ever gone while wearing that tracker can be shared. Who should we trust with the power that comes with such information?

As a society, we have repeatedly chosen convenience and financial benefit over privacy, whether it's the technology we keep on our wrists and in our pockets, the auto insurers we let monitor our driving, the apps we use for dating, the e-commerce companies we trust with our digital streaming records, the government and nongovernment entities that can follow us with increasingly ubiquitous security cameras, or the finance companies that monitor our payment histories. Much of the information we hand over for simple convenience. Debit and credit cards existed for decades before internet commerce came along. Cash was good for pretty much everything, but many people just didn't want to carry it. The desire for the convenience of using cards instead of cash gave banks and credit card companies the ability to keep meticulous records of our spending habits. We have given up our privacy, bit by bit. The digital revolution has accelerated the pace significantly, such that more than 60 percent of US adults say they don't believe it is possible to go through daily life without being tracked and monitored. More than 80 percent of us feel we have little control over the data being collected.[17]

Putting the genie back in the bottle does not seem to be an option, and it's unlikely we will forestall the collection of private data when it comes to an industry that could give us the biggest benefit of all—extra years of healthy life. The best we might hope for is effective constraints on its usage—laws that don't outright block the gathering and sharing of de-identified health data but define appropriate use and constraint.

The genome component of personal data clouds is unique to each of us. As such, it has tremendous potential to be the source of many types of discrimination. Who might act inappropriately from gaining knowledge of a person's genome? Well, insurance companies, to be certain. Employers, too. Both of these entities would stand to save a lot of money by denying access or employment to individuals whose genes indicate a propensity for expensive-to-treat diseases—and they might be tempted to charge more to insure anyone who poses a greater health risk. Even family members could use the data in nefarious ways; it's not hard to imagine that a genetic predisposition to mental health challenges could be a factor in child custody proceedings.

These are the sorts of concerns that were on the minds of US lawmakers soon after researchers around the world put the finishing touches on the first human genome sequence in 2003. In response, Congress passed the Genetic Information Nondiscrimination Act (GINA) of 2008 by a combined vote of 509 to 1. (The lone dissenter, libertarian Sen. Ron Paul, explained that he agreed with the overall goals of the bill—he just didn't believe the federal government had a good history when it came to protecting personal information.)[18] When President George W. Bush signed the bill into law, it brought to life a piece of legislation that Ted Kennedy called "the first major new civil rights bill of the new century."

It's easy to wag fingers at hypothetical violations of privacy, but there are plenty of scenarios that could potentially challenge anyone's ethics, particularly when it comes to weighing the privacy of individuals over the safety of their communities. If you were the principal of an elementary school, the manager of an elder care center, or the director of a hospital in the midst of a fast-spreading and deadly pandemic, would you want to know if one of your employees had a genetic variant that made them more likely to catch, carry, or spread the disease? Even if you can say for certain that you would be able to resist the temptation to look at the data if you could easily access it, how do you think most other people might act in the same situation?

In our vastly interconnected world, it is imperative for responsible people to game-plan for such eventualities. If revoking some of GINA's provisions would have saved millions of lives during the COVID-19 pandemic, would lawmakers have done so? Should there be exceptions to de-identification, such as the tracking of infected or exposed individuals? On the other hand, the analyses of these de-identified data may provide critical clues as to more effective diagnosis and treatment. These are the questions we need to ask and seek to answer long before the next global crisis comes along.

Even if GINA remains on the books for a long time to come, and even after several revisions aimed at strengthening prohibitions against genetic discrimination, many critics have argued that it doesn't go far enough. It does not cover disability or long-term care insurance. It doesn't apply to the military or to employers with fewer than fifteen employees (that's about 80 percent of all businesses in the United States). And it is only a US law. Europe generally has stronger laws when it comes to genetic discrimination and privacy, but Australia and most nations in Asia and South America offer few protections, and most of the nations in the Middle East and Africa offer almost none at all.[19]

In time, the massive personal data clouds at the heart of scientific wellness will include data that arise not just from personal measurements but also from the people around us. It is already clear that family-disease histories are extremely beneficial for physicians who want to know what they should be looking for in their patients. That will be even more true as we move toward P4 medicine. What may be less intuitive is that the data from coworkers, housemates, and friends will be valuable, too. It is not at all unreasonable to think that it may someday be valuable for a person's doctor to know whether the patient worked in an office where a specific COVID variant swept across many coworkers.

These extended data clouds will create a deeper view of each of us—at a molecular, social, and historical level—than anything that has ever existed in human history. If discrimination were the only concern, it should be enough to warrant an annual reevaluation of existing laws, led by the best legal minds in the world, with council from domain experts in appropriate categories of healthcare. But no nation on Earth is doing that—and discrimination is most certainly not the only concern.

Let's not fool ourselves that any set of laws, aimed at any number of potential concerns, will ever protect our privacy against the worst impulses of society's most opportunistic individuals and companies. It is inevitable that some new privacy concerns will be created, but we will not be making this tradeoff only to benefit our own lives. Data-driven healthcare promises to revolutionize twenty-first-century medicine for the benefit of our children, grandchildren, and all our future progeny—and hundreds of millions of people across the globe. This is our most compelling argument for sharing the data needed to achieve scientific wellness—to help improve life for all of humankind.

Those like us who support this vision should not dismiss legitimate concerns. What we are asking from one another is a willingness to allow our personal data clouds to be analyzed in order to pioneer the medicine of the future, acknowledging some risks and pitfalls—and while continually striving to address those ills in service to a greater good.

Filtering What Matters

In any field that is taking a large number of measurements, you inevitably get to the critical issue of separating actionable information or knowledge from additional data that can or should be ignored—the age-old challenge of isolating the signal from the noise. Our brains are extremely effective at filtering information. Consciousness itself is the result of this process. You could not be reading this book right now if you were consciously attentive to even a tiny fraction of the total information your brain is tracking and processing. This ability to pick what is important out of a vast trove of available information is inherent to what makes us human.

We don't have the same kind of intuition when it comes to the data that stem from genes, blood analytes, the microbiome, and digital measures of health. Any big data set has significant noise that must be identified and separated from meaningful signals. Making that happen—achieving massive reduction in "data dimensionality"—will take more than intuition. It will take machine learning, artificial intelligence, and deep databases of knowledge gleaned from well-designed scientific experiments.

Even when we apply these tools and focus the best minds on the challenge of extracting predictive patterns from molecular data, the process of identifying meaningful and actionable signals is slow and painstaking work. Patterns, after all, can be deceptive. Correlations, unrelated to causal explanations, are everywhere. What is truly causative and predictive? For many years, most of our lab's efforts—and the efforts of many others in this space—have been dedicated to answering these questions, but there's still a long way to go. Such efforts will improve substantially as we continue to build out the mechanistic frameworks for interpretation of large biological data sets, but that requires the creation of large data sets in the first place.

That's a big hurdle, because funding any scientific approach generally begins with a "prove it small, prove it bigger" model of increasing investment that, if continually successful, eventually leads to a "prove it nationally, prove it globally" scale of operations. This path is challenging when it comes to scientific wellness because creating these data sets is a colossal undertaking. Imagine your response if you were the person holding the purse strings and, at the point of "prove it small," someone said to you, "We're still not sure where all the signals and all of this noise are, but we're confident we'll be able to see it much better once we have generated more data in the future." You might be tempted to say, "It sounds like you're going into this blind!"

And you'd be right. In some ways, we are indeed going into this blind: If we want to visualize what scientific wellness might offer us in the future, we cannot simply rely on what we are capable of today—we have to use our imaginations, some common sense, a lot of computational power and technological innovation, and our trust in our capability to build a future we want. We also need to extrapolate what is possible with global data sets based on trillions of data points and quadrillions of patterns.

One example is a strikingly easy-to-see genetic link that allows clinicians to tell the difference between two types of cancer that appear to be very similar in clinical signs and morphology. Simply by looking at the relative expression of two genes in an individual's tumor, researchers can know with a high degree of accuracy whether the cancer is a gastrointestinal stromal tumor or a leiomyosarcoma.[20] This is an important distinction, as treatments for these diseases are very different. You'll rarely find such a clearly delineated diagnostic association: If Gene A is higher than Gene B, the cancer is almost certainly a gastrointestinal stromal tumor. If B is higher than A, it's most likely leiomyosarcoma.

Yet to find this very simple association, we had to input three million data points and sort through approximately one billion patterns.[21] That's a lot of data, but it pales in comparison to the scope of data we're talking about for scientific wellness. Most patterns won't turn out to be so simple—and there are major issues to deal with regarding data reproducibility and context specificity, because signatures can often change significantly in different hospitals or research settings or among different individuals. Such a test is also clearly only applicable in a very narrow setting, where other information has made it clear that the diagnosis must be one of these two cancers.

As we search for increasingly complicated associations (where, for instance, something is true if Gene A is higher than Gene B when in the presence of Microbe C, but only if Analyte D is lower than Analyte E in people for whom Digital Health Measure F is out of range for individuals of Race G), we simply cannot start with a small study population. It just does not work to "prove it small." We've got to start with larger populations and massive data sets. In other words, we have to collect vast amounts of data about people before we even have enough information to identify what we're looking for. That's an admittedly audacious proposition. But we feel certain it will catalyze a profound change in healthcare from a disease orientation to a wellness orientation. The differences in what we could conclude in the Pioneer 100 program with 108 individuals and the Arivale program with 5,000 individuals compellingly justifies this viewpoint.

Making Sure Testing Isn't a Smoke Screen

Scientific wellness is predicated on the idea that we should be extending our health span by optimizing wellness and gathering and acting on early warning signs of disease. Optimizing wellness will extend your years of healthy living. If you know someone who died of a disease that could have been treated but wasn't caught early enough, you will understand how important this is, as you will be painfully familiar with the feeling that outcomes might have been different, "if only we had caught it sooner." For obvious reasons, there has been lots of interest in early detection of cancers, which can be key to survival. Interventions that come late, especially post-metastasis, are much less likely to be effective. So it might come as no surprise that some of the staunchest critics of the scientific wellness approach to healthcare come from oncology.

In 2013, before the Pioneer 100 study had been completed, and long before any of its results were published, one cancer researcher who wasn't privy to exactly what the study was or its early indicators of success published a hostile critique in a scientific journal about our recently initiated wellness study. It would be natural to be annoyed or even angry at such an attack, but the only decent response for critics from the field of oncology is compassion. These are people who have given their lives to fighting cancers, and they've been burned again and again by the promise that early detection was a panacea—so much so that, as recently as 2019, a group of international researchers wrote that it was "time to abandon early detection cancer screening."[22]

Screening has long been a hallmark of cancer control efforts across the nation and around the world since the Pap test for cervical cancer was introduced in 1923. The decades since then have brought declines in overall cancer mortality. "However, the lion's share for this improvement seems to have been due to therapeutic advances rather than screening," the researchers wrote.

"Screening is big business: more screening means more patients, more clinical revenue to diagnostic and clinical departments, and more survivors in need of care and follow-up," the researchers continued. "We believe, however, that a major, radical change is urgently needed after more than four decades of enormous investments and failing expectations."

Even if it were true that screening was only marginally effective, would it not be a good investment given the relative costs and consequences? When cancer is not caught early, the choice that remains is to destroy or remove cancerous cells with surgery, radiation, or toxic chemotherapies—all of which cause significant long- and short-term harm and are far more expensive than routine screening. But we must acknowledge that there are significant problems with screening.

False-positive cancer screening test results are so common that many doctors have concluded that some of these tests are worthless. More than half of women who undergo annual mammography for a period of ten years will have a false positive for breast cancer, and many of those women will come to consider that experience to be the most frightening moment of their lives. About 10 percent of men undergoing regular prostate-specific antigen (PSA) testing will have a false positive experience for prostate cancer. And about a quarter of those who have regular examinations of stool samples for hidden blood will experience a positive result that *may* suggest colon cancer.[23]

The population screening challenge is highly daunting, and basic mathematics makes it easy to understand why. Consider a hypothetical blood test that showed exceptional performance, with 98 percent sensitivity and 99 percent specificity in a clinical trial with a typical 50/50 split between cases and controls. That is an excellent test result, but when applied across a population of a few million people where the expected incidence rate would be low—that is, vastly more people don't have the disease than do—there could be tens of thousands of false positives or more.

What's even more alarming is when people who receive false positives go on to receive treatment that is completely unnecessary.[24] Most treatments for cancers—including chemotherapy, surgery, and radiation—are toxic, invasive, and expensive. This has left a bad taste in the mouths of many cancer researchers, so much so that the very notion of early detection through biomarker screening has, for some, become anathema. We understand and even agree in some cases, due to high false positives and the major downside of the interventions pursued.

Not all diseases require the sort of drastic interventions that cancer treatment often demands, and extending our health span certainly doesn't. In scientific wellness, the preferred interventions are those where the tradeoff between cost and benefit for false positives is far more favorable than what we see in fields like oncology. The sorts of lifestyle interventions that are most effective for people with prediabetes, for instance, are overwhelmingly safe, involving sensible changes such as lowering sugar intake, being more active, and increasing fiber intake. Supplemental use of a probiotic with *Akkermansia muciniphila*, a beneficial species in the human intestinal microbiome, has been shown in an initial multisite, randomized, controlled trial to improve glucose control.[25] These interventions are almost always holistically beneficial for general health, though to differing degrees. Even when people are unable, or unwilling, to go through with these sorts of interventions, they can still usually prevent or delay the onset of later stages of disease by improving metabolic health markers with natural products such as berberine or small doses of particularly safe medications such as metformin.[26]

It's also important to note that large numbers of measurements enable different strategies for error correction. In a sufficiently large data set, a single outlier in a biomarker can be judged by whether assemblies of other corroborating changes are seen across the other measurement data. This can't be done with single or small numbers of biomarkers, but it's relatively easy to do when lots of measurements are being taken and there is more than one indicator of a wellness or a disease transition. It's one thing if a series of tests results in the identification of one blood analyte for which a shift in values has been associated with a departure from wellness. It's a very different thing, though, if testing also reveals a change in other analytes that have been independently associated with the same condition—and more so if the individual is genetically predisposed to this condition, and even more so if there has been a change in the microbiome that often correlates with this transition or disease. With each additional piece of information, the window for false positives narrows. It is for this very reason that we believe a greater sensitivity and specificity in monitoring for a wide range of diseases will be most effective with the deep phenotyping (and the generation of lots of data points) of large numbers of individuals.

Patients Want More Data

Outside the cancer field, some doctors worry that too much data will increase anxiety, turning people into anxious hypochondriacs. This potential problem is exacerbated by an explosion of information online (some of it credible, much of it not) that encourages people to self-diagnose (occasionally accurately, many times not), connecting their symptoms to all manner of rare and deadly diseases and self-medicating using spurious and potentially toxic supplements from low-quality sources, sometimes with terrible consequences.

In both the case of conventional cancer screening and that of "scary information overload," the perceived problem is the same: when you apply simplistic diagnostic criteria across a large general population, you are certain to have plenty of false positives, inviting fear, overtreatment, unnecessary expense, and diagnostic skepticism. The solution is also the same—more data. This is why, although screening resistance remains quite entrenched, we are starting to see the pendulum swing as we increasingly validate the power of a deep data-driven approach to healthcare.

Take the case of the simple test we developed at ISB to identify a gastrointestinal stromal tumor versus a leiomyosarcoma. This test is of no benefit to the entire population. Everyone in the world has a greater expression of one gene than the other; that doesn't mean everyone in the world has one cancer or the other. The false positive rate of such a test in a large population would be enormous. It's not until we actually have reason to believe that someone has one of these two cancers that the simple diagnostic tool becomes useful.

What makes any test valuable is other tests, because additional tools of analysis provide a richer context for interpretation. The more background criteria (data), the better. A person with a genetic predisposition to a disease is a person who will most benefit from a blood analyte test that can signal that a transition to that disease is under way, because disease reversal can begin immediately. These are also the individuals who will benefit the most from screening tests that help determine the courses of prevention and intervention most likely to help them. And finally, these are the individuals who have a reason to be attentive to what specific screening tests might indicate. People in such situations are not suffering from hypochondria; they are exercising a vigilance that could end up improving their health and even saving their lives.

For most people, identification of high genetic risk, early diagnosis, and frequent widespread testing is not nearly as scary as some in the medical community have long assumed.[27] Even before the Human Genome Project was completed, bioethicists feared that people who learned they had a genetic propensity for disease would be overwhelmed with anxiety. That's a concern that has persisted despite the fact that research doesn't back it up.[28] Studies have shown that individuals who choose to learn of their genetic risk of developing Alzheimer's disease do not experience any more negative psychological impacts for having received that news.[29] And as we will see in Chapters 7 and 8, there are concrete approaches one can consider to prevent the transition to Alzheimer's disease.

The key really is data density and context. Denser measurements provide much more information to correct false positives than any single test. Anyone who had a hard time recognizing a mask-wearing friend during the COVID-19 pandemic knows this to be true: it is hard to recognize someone from seeing just their eyes, but when you combine their height, hair, body shape, and voice, an identifiable picture emerges.

We came to see this at Arivale. The more data participants had that were specific to themselves, the better they felt—even if a singular piece of that picture was potentially "bad news." If they were predisposed to a condition, they had peace of mind knowing that, if a disease transition were ever initiated, it would likely show up somewhere in their data and could be dealt with early. It wasn't until Arivale failed economically that many people expressed anxiety, as they no longer had a sentinel-health system in place, standing watch over optimizing their wellness and identifying their disease transitions.

We don't have to test everyone for everything. We need inexpensive and broad testing to identify individuals at highest risk, and then we can take more precise measures of those individuals to home in on solutions to early transitions toward disease. Until we can do that effectively, we will not be able to prevent unwarranted fears or unnecessary treatments, or fully address the valid concerns many people have about false positives. But with each new data-rich study, the accuracy and effectiveness of data-driven health improve. Critically, interpreting information from multiple layers of different data types simultaneously is essential for dramatically improving diagnostic accuracy in the long run.

Bringing Down the Costs

Arivale's demise came to pass, in part, because of the costs of the assays needed to supply the genome and phenome data. While these costs were already falling during Arivale's brief but productive existence, they have fallen even further since then and will continue to do so. Many people have noted that, while the first human genome cost nearly $1 billion, the price for an even more detailed human genome sequence today is just a few hundred dollars, and it will eventually be so cheap, and so valuable, that insurers will be forced to make genome sequencing a standard feature of any individual's healthcare plan. As of this writing, you can get a whole genome sequence done for a few hundred dollars by companies such as Nebula Genomics and Dante Labs. The cost for measurements of dynamic blood analytes will fall as well. The costs associated with assessing our microbiomes and maintaining frequent measurements of digital health will also continue to drop; we've seen that in our own work with Thorne's new "microbiome wipe" technology.[30] Finally, the million-person project for genome/phenome analysis proposed by Lee, which we will describe in greater detail in Chapter 11, if successfully funded, should help drive down the cost of phenomic analyses by orders of magnitude, from thousands of dollars to tens of dollars over the next ten years or so.

Victory at last? Hardly. Assays are just one of the costs with which we must contend. There are also costs associated with developing and deploying the computational algorithms needed to characterize each individual's health status, carrying out the recommended actions, and continuing to develop the needed health intelligence (research of this sort is expensive), as well as the costs to deliver care via the doctors, coaches, and other experts who will treat and guide patients.

There are other costs at play, too, and they are perhaps the most powerful economic forces acting against the data-driven health revolution. These are the infrastructural, organizational, and personnel costs that have been incurred by current healthcare organizations, insurers, research organizations, pharmaceutical corporations, and countless other interests that have trillions of dollars invested in the way things are now. These costs capture 17 percent of the US gross domestic product every year. That's a tidy sum. As such, it should come as no surprise that there are enormous vested interests that insist on clinging to the status quo.

At risk are the mortgages for hospital buildings, surgical suites, and expensive technological equipment that could be made obsolete by a shift to scientific wellness at home, as well as investment in decades-long pharmaceutical studies that could be much less profitable if disease transitions are identified early and reversed. Trillions of dollars have already been spent for infrastructure and disease strategies that are expected to pay off in the long run. These investments were made in a world in which medicine is practiced a certain way. If that changes, the equation changes. And that's a scary proposition for a lot of people.

Any decent poker player will tell you that you cannot make decisions based on sunk costs. Most will also tell you it is hard to abandon a pot in which they have heavily invested. This, of course, assumes that the player has their own best interests in mind—and that they are playing with their own money—but that's not the case with many entities invested in the healthcare status quo. They're playing with *your* money—and *your* life. This does not mean that those who are invested in the current system don't care about what is best for the patients who rely on the system. There are many different ways to think about well-being, and while maintaining the status quo isn't always the wrong move in the short run, it is frequently the wrong choice in the long run.

In the United States, the healthcare sector employs more than a tenth of all workers and accounts for a quarter of all government spending, while health insurance accounts for a huge part of working compensation, all of which has led economists to call the healthcare sector "the most consequential part of the United States economy."[31] A sudden shift in the foundations that support this sector would send shock waves through the system the likes of which have not been seen since the years following the Flexner report, which prompted the closure of more than half of all medical schools.[32] There will be collateral damage, some of which we can predict and some of which we cannot.

It is only when we appreciate the totality of these costs that we can consider the savings that will ultimately result from health improvement and disease reduction. Even then, we must remember that "saving money" isn't the goal: optimizing and saving lives is. Making that goal our North Star takes the power away from those who would use the easily manipulated concept of "cost and spending reductions" to maintain the status quo and protect those who most profit from it.

Ultimately, scientific wellness and healthy aging will provide enormous cost savings—and these savings will largely go to the payers. This is precisely why healthcare systems should become integrated payers and providers—the cost of innovation can then be shared, and the savings can, too. But savings on the part of providers are only worthwhile to the rest of us if at least part of what is saved is reinvested in initiatives that provide benefits in terms of extended years of healthy life.

It is already happening. Cost cutting is one of the key measures by which corporate boards decide how they'll reward healthcare chief executives. And make no mistake: the nation's top executives are being well rewarded. In 2018, the CEOs of the top 100 healthcare companies made $2.6 billion.[33] That's about a half-billion dollars more than the National Institutes of Health spent researching Alzheimer's disease that year.[34] This is where "value-based healthcare" comes in—a paradigm in which healthcare systems get paid for keeping people well rather than for how many times they see a patient, as is the principle of the current healthcare system. The *value-based healthcare* imperative will drive the system to scientific wellness and healthy aging approaches to ensure healthy, and hence cost-saving, patient populations.

So how do we get there?

Extending a person's health span might not save a lot of money in the long run because, at least under the current healthcare paradigm, the savings associated with a person needing fewer health interventions each year will be offset by the costs associated with their needing strikingly fewer interventions over a greater number of years. And even healthy people do get sick eventually. The hope for scientific wellness and healthy aging is that most of us will move into our nineties mentally alert and physically healthy, needing few, if any, medical interventions throughout most of our lives. Once you get to 100, most people die rapidly by a sudden systems failure without protracted and expensive healthcare costs. If this optimistic view becomes a reality, there will be enormous savings for healthcare and for families that are often burdened with enormous medical costs in the last years of their loved one's lives.

All of this complexity aside, can we say with some degree of certainty that the end result of a wellness-driven model of healthcare will buy "more health for fewer bucks"? After all, every year of a healthy life has serious value both in savings and productivity—not to mention happiness, which is more important but harder to quantify. Research shows that prevention is a lot cheaper than treatment, a much more cost-effective approach to "buy health," particularly if it achieves the goal of preventing diseases that last a long time and are expensive to treat and care for, such as Alzheimer's, diabetes, cancers, and cardiovascular diseases.

So how do we simplify our approach to monitor for health across a broad spectrum of metrics and quantify individual improvements or declines? A good place to start is by understanding the principle of biological age.

5

A New Way to Think about How Old We Are

Why Biological Aging Is an Essential Concept for Lifelong Wellness

For many children, birthdays are a singularly fantastic time. For one amazing day, they experience the feeling of being at the center of the universe. They have a sense that they are growing bigger, stronger, smarter, and older. As we grow up, birthday excitement gives way to trepidation. Our health begins to decline. We are no longer hopeful about growing older but wistful for our youth. But are we yearning for fewer years or fewer of the ailments that come along with those years?

This vital distinction is the difference between age and aging. The first is marked by trips around the sun. The other is a biological process marked by a progressive accumulation of damage over time, leading to impaired function and increased vulnerability. For most of human history, age and aging have been almost perfectly linked. We take aging for granted—so much so that we've done very little to fight it. But while growing older is inevitable, aging isn't. In our view—and that of an increasing number of researchers—it is a condition that can and should be fought.

Aging is the number one risk factor for virtually every major chronic disease—and many infectious diseases as well, as we saw in the case of COVID-19, which was far more dangerous for older people than for the young. Age was such an overwhelming risk factor, in fact, that Vadim Gladyshev of Harvard Medical School wrote a paper arguing that we should consider COVID-19 to be a disease of aging.[1] But the coronavirus is not alone in this regard. In response to almost every disease, young people are far more resilient than older people. It isn't a tremendous leap to argue that the aging process itself is central to the onset and increased risk of many diseases, both chronic and acute. And it stands to reason that if we could optimize our wellness through healthy aging, slowing down or even reducing our biological age, we might be able to fight these diseases as effectively as we could when we were younger. Imagine what a difference that would make!

Although the proposition is controversial, we have come to think of aging as a "disease in slow motion." If a hypothetical disease did in a matter of months or years what aging does to you over a lifetime—brittling bones, wasting muscles, fogging cognition—we would consider it a terrible scourge. But the generally slow pace of decline is deceptive, as one gradually becomes accustomed to the ever-increasing compromises that come with growing older. Let's be clear: reversing the effects of aging is a tall order. However, exciting advances in just the past few years have infused hope in this once impossible dream, including multibillion-dollar investments in regeneration and longevity companies such as Altos Labs and Calico. Our bodies have tremendous regenerative powers, but over time, many sources of damage—to DNA, metabolic enzymes, proteins, bones, muscles, and more—begin to accumulate. Eventually, the information carried by our genes and its physical manifestation begins to decay. Without regeneration, entropy wins. We age and die.

But at what pace shall we permit this to happen? This is one of the most important questions of our time. There is no fundamental law of biology that suggests aging must happen at any given rate. In fact, evidence strongly suggests there is no universal rate of aging: different mammalian species age at strikingly different rates—some living 200 years and perhaps longer.[2] There was even a species of shark recently discovered to live nearly 400 years.[3]

Just about all of us know 50-year-olds who, by virtue of lifestyle choices or genetic misfortune, look older than their age. And most of us also know people who, because they made good choices in life or got lucky in the genetic lottery, are part of the growing ranks of the "young-old," those who reach their eighties with all the vigor, mental acuity, and healthfulness they had decades earlier. Will the young-old push human life spans beyond the brink of what we now think possible? To 120 or beyond? We think that could happen. But there is a far more pressing concern for almost every person on this planet. Long life spans are almost impossible, and certainly unpalatable, without long health spans. Our life spans are the length of time we live, irrespective of how we live; our health spans are the portion of our lives spent in good mental and physical health. If you explain this distinction to any group of people, anywhere in the world, you would be hard-pressed to find any concept that garners greater agreement across social, cultural, religious, political, and economic barriers. Not everyone wants to live longer, but just about everyone wants to live healthy for longer. Health spans are more important to most of us than life spans.

That's good news, for while extending lives past the known extreme of about 120 years is a tremendous challenge, slowing aging to extend health span is more readily attainable. Despite the paucity of research in this arena, scientists have already identified many ways to slow biological aging, which has the potential to add more years to our health spans than anything else. It means more than curing cancer, heart disease, respiratory disease, or stroke. Conventional estimates of the years that could be gained if we were able to eliminate any one of these diseases are often vastly overestimated.[4] Why? Because these approximations typically ignore the fact that most people, as they age, acquire more and more chronic diseases, and eliminating any one of them does not meaningfully make the others less likely.[5] Even if we were to eliminate all cancers—every single one of them—researchers have calculated that we would only see a three-year increase in life span, and it might not affect the average health span even that much because other chronic diseases would be waiting in the wings.[6] This doesn't make the fight against cancer unimportant, but it does suggest that a disease-by-disease elimination strategy offers diminishing returns. By contrast, slowing aging, even by a little, would offer disproportionate benefits by delaying all chronic diseases.

The idea that aging is a disease is not new. That concept, popular among classical philosophers, was already around by the time Galen of Pergamon came along in the second century. Most thought aging was inevitable, but Galen, a physician, surgeon, and prodigious writer of medical texts, wasn't convinced that aging was a condition for which there was no treatment. In his treatise on health, *Hygiene*, he suggested that aging could be eased and delayed, and that any physician worthy of Hippocrates' long shadow should not only be attentive to the needs of elders but to the moral mandate of helping to prevent aging whenever possible.[7] Galen himself is thought to have lived into his eighties, healthy in body and mind for most of his long life.

Over the years, there have been many theories about why we age. Could aging simply be a result of the exhaustion of the finite internal and external resources available to a specific member of any given species? Could DNA damage and accumulated mutations lead to a loss of genetic information? Perhaps it is related to the oxygenation of DNA by unpaired electrons. All these hypotheses and more have been explored. Some have accumulated a fair amount of supporting evidence. But none has done much to shift the conventional view that aging is inevitable.

More recently, researchers have suggested that aging may be the result of a combination of factors, including changes in the ways genes are expressed over time, a growing inability to get rid of dysfunctional proteins through autophagy (engulfing proteins by macrophages—a type of white blood cell), the exhaustion of stem cells, or the accumulation of zombie-like senescent cells that inflame healthy cells.[8] Together, these are referred to as the "hallmarks of aging."[9] In this way of thinking, each of these factors has the power to perturb the others.

Is this starting to seem to you like a systems challenge? It is indeed, and that's the kind of challenge we find irresistible. Over time, it was becoming clear that one promise for changing the paradigm for how we approach healing and health would be to identify a biological system or systems upstream of these aging hallmarks that control them. Even if we found such a system, we would have to find a way to manipulate it to disrupt downstream consequences.

Back in 1993, biologist Cynthia Kenyon provided compelling evidence that such identification might be possible when she published a startling study showing that a mutation in *a single gene* could *double* the life span of a roundworm known as *Caenorhabditis elegans*, and that a second different mutation could reverse this improvement. The idea that there might be "on and off switches" for aging inspired an arms race in the field of aging research, leading to many discoveries of similar triggers in other organisms, including mammals like us.

Why would such triggers exist? Harvard Medical School geneticist David Sinclair has proposed an intriguing possibility.[10] Sinclair, who first came to international prominence when he published a study suggesting that certain chemicals could extend the life span of yeast and other simple organisms, and whose work later showed how to generate significant life span and health span improvements in mice, has theorized that all life-forms share a common ancestor that developed "survival genes," which function to shut down DNA replication when times are difficult and which promote functions that facilitate reproduction when times are more favorable.[11] Shutting down DNA replication slows aging by turning off one or more of the hallmarks of aging. More than two dozen such survival or "longevity" genes have been discovered in humans, and several have been deeply studied. Evidence is mounting that these genes are at the heart of a system—a surveillance network, if you will—that releases proteins and chemicals into the bloodstream in response to sleep, what we eat, how we exercise, and even the time of day. These genes have been primed by billions of years of evolutionary history to "hunker down" when times are tough. Conversely, they respond to a relative lack of stress—bountiful food, a lack of exercise, temperature-controlled environments—by facilitating cellular replication—and with it, aging. Sinclair believes that exposing our bodies to healthy stress in the form of intermittent fasting, rigorous exercise, and cold temperatures may compel these genes to conserve energy, slowing the replication of cells and blocking many, if not all, of the hallmarks of aging. This insight could be a starting place for major innovations in human wellness, as we turn our attention to upstream causes of aging rather than downstream consequences.

As tough a problem as aging is—and as closely integrated with ordinary human life as it may seem—we now have some of the tools we need to slow, stop, and even reverse it. As these tools improve in the coming years, the result will be nothing short of world-altering, as we embrace an idea of wellness that doesn't just mean aging gracefully but aging much more slowly, if at all. But as is often said, you can't hope to have an effect on something you can't measure. So that's where we need to begin—we need to find a metric that assesses aging.

Finding the Right Way to Measure Aging

Aging is a gradual process. You can't see it advancing day to day, week to week, or month to month. Sometimes it's hard to see from year to year. For the most part, its effects accumulate slowly; we get an extra gray hair here, find a deeper wrinkle there, get a little less flexible. Sometimes the effects wax and wane: if a brain beginning to experience the effects of aging rewires itself ever so slightly, we may feel renewed clarity of thought, perhaps for years to come. This makes aging hard to measure. But even if we can't measure it precisely as we would the size of a tumor, emerging research has given us tools to quantify it.

A quantifiable metric might best be thought of as a global assessment of life processes that span the whole organism as well as those operating in specific systems or organs. Wellness doesn't just mean an absence of disease but a resilience to future ailments and the energy for activities that enrich our lives. Aging is the opposite. Measuring health spans may look like a good start, but health spans are only really measurable once a person reaches a persistent state of disease. Even if it's an easy number to calculate, it's not a particularly actionable metric because wellness-to-disease transitions can happen years before symptomatic morbidity.

Perhaps counterintuitively, the measurement we really need is more akin to chronological age—a number that will give us a sense of where a person stands in the aging process. We know that in the vast majority of cases, a person who is 60 is further along in the aging process than one who is 40, so age can certainly be a generalized stand-in for aging. The problem is that it becomes harder to use age as a proxy for people who are 60 and 62. Who is more likely to be less well? Given two completely random individuals of those ages, would you be willing to bet a large sum of money on the fact that the 60-year-old is healthier than the 62-year-old? That wouldn't be a good bet. And this gets to the heart of three problems with age as a metric for aging. First, it's not very precise. Second, it doesn't capture information about how well or poorly you have lived your life. And, finally, it is not modifiable—that is, it does not lead to actionable possibilities for change. It is, in a way, like using an odometer and nothing else to determine the value of a used car. Would you pay someone $5,000 for a vehicle with 150,000 miles on it, knowing absolutely nothing else about the car? Likely not. Five thousand dollars for a car could be a bargain or a rip-off, and mileage is only one of many factors you would want to consider. At a minimum, you would also want to know the make, model, and year, as well as the record of ownership, the history of accidents and repairs, and the experience of other owners with a similar car. The same is true for anyone who wishes to understand wellness in the human body.

Therein lies the challenge. Chronological years are a murky reflection of actual biological aging, but biological aging can be an intuitive concept to grasp if it is mapped against chronological age year by year. So how do we bridge that gap? We used an algorithm that forced the slope of its biological aging estimate to be near one, such that one unit of biological aging was comparable to one unit of chronological age.[12] This calculation is essential to the most important use of biological aging—to track it repeatedly in your own life so as to know your rate of aging and provide an action plan for how best to optimize your efforts to live your healthiest life.

For simplicity, we could denote someone's chronological age as "a" and their biological age as "b." An exceptionally average person who is 60a is also just about 60b. Someone who is 45a could be 60b, too, indicating faster aging. And someone who is 75a could also be 60b, indicating slower aging.

When you follow these changes over time, the rate of annual change is meaningful. If today you are 50b and next year you are 52b, that's bad news; you are experiencing biological aging at a significantly accelerated rate. And if today you are 50b and next year you are 51b, that's perhaps to be expected, but it's also not great news; you have still experienced typical aging. But if today you are 50b and next year you are still 50b, that's quite good, because you have not increased your biological age over the year. And if today you are 50b and next year you are 48b, that's terrific; you have reduced your biological age by two years!

Being able to see the cumulative effects of efforts to reduce biological age over multiple years is another benefit to such a metric. It allows one to calculate a "delta age"—the difference between chronological age and biological aging. If you've had fifty birthdays but have a biological aging mark of 55, your delta age is +5. On the other hand, if you've lived for fifty years but an assessment of biological aging pegs you at 45, you have a delta age of –5. Obviously, this is one of those times when it's good to be negative.

Many metrics for biological aging are emerging, and you can compile a delta age using any of them. We believe that the metrics that will be most accurate will be those based on more measurements and enormous cohorts. An exciting discovery that emerged from the data clouds of the 5,000 Arivale customers was a new multiomic approach to determine an individual's biological age. The development of this metric was spearheaded by John Earls, a computer science graduate student and software engineer in our lab. This algorithm offered the most comprehensive assessment of aging across many biological systems because it was derived from a large population of people from their early twenties to over 90, integrating whole-genome sequencing plus longitudinal blood measures of more than 850 proteins and metabolites. The approach enabled us to assess biological aging across a very large number of different biological systems.[13]

The result was not just a new way to assess each person's biological and delta ages; it showed the relative effect of different diseases on biological aging. Of the chronic diseases we investigated, diabetes had the largest consistent effect on raising biological age, with an average of over six years. Smoking resulted in a biological age of about two years higher in this analysis. The primary drivers of biological age across all participants were metabolic health, inflammation, and the accumulation of biotoxins. John's algorithm generated a host of actionable possibilities for improving biological aging by addressing these problems uniquely for each individual.

There may be no better way to understand holistic health—and optimize wellness—than to have a handle on your biological age and track it over time. Fortunately, researchers across the globe are continuing to develop increasingly accurate metrics that, like the one that came out of Arivale, provide a window into whole-body biological aging as well as the aging of individual systems, as research shows that different organs age at different rates.

No matter what method for assessing biological aging is ultimately adopted by the greatest number of people, it is important that we quickly come to a place where we can easily use these metrics to track our health and improve it. With a universally adopted metric, we can start to imagine a program to reduce biological age across a population and its tremendous impact in terms of both extended health spans and reduced healthcare costs.

A lot of resistance to wellness can be summed up in the words "aging is inevitable." When people can see that the rate of aging is not immutable—when they can follow a number that can go up and down in response to what they eat, how much they exercise and their stress, toxins, and sleep—it will help promote healthy aging. People are free agents, and not everyone will choose to act on this information, just as many smokers today don't have the inclination or willpower to quit. But the choice, at least, will be in their hands. We also must take collective actions to make personal choices toward wellness as easy as possible through access to wellness-centric healthcare, broad availability of healthy and delicious foods, and influencing social norms toward routine acceptance of healthy behaviors such as lessening our time sitting in offices, eating less junk food, and much more.

What Measurements of Aging Offer

Can a person's delta age really impact their overall health? Does it give us a substantially better measure of health than chronological age? And does it help create actionable opportunities for improving wellness? The answer to all three questions is yes.

Using data from Arivale, we measured whether a person's delta age was positive or negative and then compared the disease history of those groups. In analyzing about forty different diseases, the result was an across-the-board correlation between higher deltas and a history of disease. As previously noted, the most dramatic of these correlations came among people with type 2 diabetes, which resulted in an average delta age of +6 years. This aligns to findings from researchers at the Centers for Disease Control and Prevention, among others, who have noted that "the anatomical and physiological changes characteristic of the normal aging process are accelerated in people with diabetes."[14]

So does aging cause disease, or does disease cause aging? We believe it's both. But either way, this much is clear: keeping your delta age as negative as possible is a key factor for decreasing your chances of disease and increasing your health span. And the good news is this is something most people can control to an impressive extent.

The way we traditionally think of aging would lead us to believe that a person's delta age would remain more or less constant over time. Thus, whenever we add a year of life, we add a year of biological aging. By this view, the natural mean state is a delta age of 0—and this was built into the algorithm we used to derive our version of a biological age estimate. But the people enrolled in the Arivale scientific wellness program completely shattered this notion. Instead of seeing a year-to-year increase in biological age, we saw an annual decrease in delta age of about 1.16 years. That might not seem like much, but consider what it really means: maintained over time, these folks were expanding their negative deltas. They were maintaining—even slightly reducing—their biological age!

For the women who enrolled in Arivale, the effect was particularly dramatic, with an improvement in delta age of 1.5 per year in the program. Men saw an increase of 0.2 units of biological age per year, much less than the 1 year expected. They thus had an improvement in delta age of 0.8 per year. They were on pace to live five years but only experience a one year increase in biological aging by this metric.

Nathan achieved a striking result over the four years he participated in the Arivale program. During that time, his delta age improved by an average of 2.7 each year, resulting in a biological age that was ten years below his chronological age at the end. He did his best to follow the advice of his Arivale health coach, Jessica Roberts, who checked in with him monthly throughout the process to course-correct and make recommendations. Changes included not replacing his car after it broke down in the mountains and walking to and from work each day (almost five miles round trip, which he came to love). He would also take calls outside on headphones and do "walk-and-talk" meetings instead of sitting, when possible. He used a Fitbit to see daily progress and track steps. He watched his diet more and monitored his blood to address nutritional deficiencies such as low vitamin D and omega-3s, which had fallen sharply to a very low level and were eventually brought up to normal levels through heavier supplementation. One thing he did that was uncommon in the Arivale population was take nicotinamide riboside (NR), which boosts NAD^+ levels in humans and has been tied to an increased health span in a number of animal models (more on this below). It isn't possible to know how much of an effect this may have had on Nathan being such an outlier in reducing biological age. (The relationship between NR and biological aging in people is something we will be evaluating more closely in a large population going forward.)

The delta age metric gives us an easy way to visualize and communicate the impact of various lifestyles. How negative could a person go? It is certainly true that, as people's biological age gets lower than their chronological age, further decreases become more difficult, but for now, just how much one can reverse the clock is an unanswered question.

It's not a question we would even have thought to ask—let alone sought to answer—just a few years ago. Does this hint that aging might be not just slowed but actually reversed? We don't know for sure, but there are fascinating studies of this capability in animal models.[15]

Thanks to our data and other studies, we can now say with great certainty that biological age increases with disease and decreases with healthy behaviors.[16] Indeed, using data from the National Health and Nutrition Examination Survey (NHANES IV) of over 11,000 individuals, Yale professor Morgan Levine and colleagues showed that biological age was more predictive of life span (all-cause mortality) and health outcomes than chronological age.[17] This finding was robust across different ages, ethnicities, levels of education, health behaviors, and causes of death. It was a remarkable result. Biological age isn't fixed to chronological age—at least not in the one-for-one way we've been led to believe. This is a destiny-shattering idea—one that opens many fascinating questions.

What is a biological age that represents a state of ideal wellness? Presumably, that number lies somewhere in the twenties—a point in chronological age when most adults have not experienced much damage to the biological systems that sustain their existences, have full function of their bodies, are resilient to myriad ailments, and have plenty of available energy. (Life satisfaction is another wellness indicator, but at least for now we'll leave to psychologists the question of whether 20-year-olds are more or less satisfied with their lives.)[18]

Can we maintain a biological age far below our chronological age over decades? Is there a limit to a person's delta age? Lee, a fanatic for exercise, has a biological age that is fifteen years below his chronological age, so we do know that the two ages can be quite diverse. Obviously, at any point of adulthood, we cannot expect to "un-age" ourselves into childhood, like F. Scott Fitzgerald's backward-aging Benjamin Button. But could someone who is 40a have a delta of –20? And twenty years later, could that person still be 20b, such that their delta would reach –40? Can we imagine a 60-year-old whose biological systems all look and act like those of a 20-year-old? We can.

As we consider such questions, we should remember that not all physiological changes associated with aging in adults are bad. Just because something changes with age doesn't necessarily mean we should strive to "fix" it. In fact, some of the changes that we experience, particularly at the biochemical level, may be protective compensatory mechanisms for other issues that come about as a result of simply being on planet Earth for many decades. We're all going to get bumps, bruises, broken bones, and a big helping of bacteria along the way. The sun will radiate Earth and everything on it. The fact that our bodies shift in response to the collective insult of these individual factors can be useful and protective.

It should go without saying that cosmetic changes—facelifts, tummy tucks, hair transplants, and such—don't do anything for your delta age. But there are lots of areas that most certainly can be impacted, because a broad base of measures gives broad insights into actions you can take. If your cholesterol is bringing up your delta age, there are actions you can take to reduce it. If your white blood cell count is contributing to a situation in which your "a" and "b" ages are getting a little too close for comfort, there are steps you can take to address that, too.

The most crucial point is that simply knowing that you can take actions to slow aging—and can see the results—is the fundamental shift most people need to begin a holistic effort aimed at improved wellness. So what actions let us age in a healthy manner?

Taking Action on Aging

Sometimes aging discussions have a futuristic quality to them, and it is true that we're a long way away from being able to say that we understand all the factors that lead humans to age or the interventions that can slow, stop, or reverse that once seemingly inevitable state. But there are plenty of steps you can take right now to leverage what is known. Several tests of biological age are already on the market, one of which we developed (Thorne's Biological Age) based on clinical labs in which we aimed to maximize health relevance and actionability around the most essential measures. Other biological age tests are available through companies such as AgelessRx and Elysium. The science is still developing, and it will take some time before rigorous studies will be able to give us a better look at how—or whether—acting specifically on these tests works on an individual basis to extend life spans and health spans. For now, commercial tests from reputable companies that focus on behaviors and biomarker changes known to benefit health are a good starting point.

University of California Los Angeles geneticist Steve Horvath pioneered the first approach to a metric for biological aging using epigenetics. Epigenetics measures the extent of methylation of one's DNA and hence the extent to which gene expression is blocked (thus slowing aging). While epigenetics is a complicated and fast-moving field, the "Horvath clock" is a simple tool—an analysis of hundreds of epigenetic chemical markers that accumulate on an individual's DNA (specifically the addition or subtraction of a methyl group to the cytosine nucleotide when it is followed by a guanine nucleotide) and change the activity of the genome (e.g., its ability to express genes) without changing its sequence. These markers are relatively uncommon in youth but accumulate over time in response to biological stressors at a pace that is rather predictable. The Horvath epigenetic clock has been shown to be a good predictor of chronological age within a few years.[19]

Being able to predict someone's birthdays by looking at the chemical markers on their DNA is a nifty trick, but its real value lies in the degree to which the delta age gives us meaningful health insights. People who are quantifiably more well—those who eat healthier diets, maintain physical activity, and avoid chronic stress, toxins, and obesity, for instance—have fewer negative markers.[20] Their clocks appear to be running slower or they are aging more slowly. Some studies have suggested there are associations between this epigenetic clock, disease, and mortality, although this remains an emerging area of research and more study is needed.[21] In fact, the newness of epigenetics makes its practical implementation tougher because relatively little is known about how epigenetic markers could lead to personalized health recommendations to improve healthy aging. This might be one of the reasons medical systems have not (yet) coalesced around methylation clocks as an age-replacing approach to quantify aging—the complex sum of damage, function, resilience, and available energy that make up biological aging. These and related metrics are continuously evolving, with increasing ties to health relevance, so we believe it is only a matter of time before they hit common usage.

Journalist Ainsley Harris's experience represents another reason epigenetic aging hasn't yet been embraced by the masses. When the 36-year-old Fast Company reporter took a test for a story, she learned that she had a biological age of 34. "It's about where I expected," she wrote. "I've always exercised, and I generally eat well. But with a 1-year-old son at home, I don't have time to be fanatical about either of those factors, or about my sleep."

How would Harris feel if she'd paid $500 to learn that her "a" and "b" numbers were close to one another, just as she had expected? "I would likely be disappointed," she wrote, adding, "I haven't learned very much about myself, or about my body's performance."[22]

That's not an unreasonable reaction for a person in their mid-thirties. We think the key is to provide not just a score but a plan of individual action for healthy aging. Learning your biological age is interesting, sure, but what if that knowledge comes with an actionable plan to optimize your personal aging and health span throughout your life? That's exciting.

Biological age assessments made from clinical lab tests give us clear and well-validated actions to take to improve healthy aging. To this end, we developed our initial consumer offering: Thorne's Biological Age. We became acquainted with Thorne first through Nathan's longtime friendship with Joel Dudley, who was executive vice president for precision medicine at Mt. Sinai in New York and cofounder of Onegevity, a start-up half-owned by Thorne. The development of this biological age test was our first interaction with Thorne. In the process of working together, Nathan received an offer and ultimately joined as CEO of Onegevity and then chief scientific officer of Thorne HealthTech following the merger of Thorne and Onegevity. Biological Age assesses both a global delta age and the delta ages of different organ systems and biological processes based on the following information:

- Blood and immune system measurements
- White and red blood cell counts and their relative levels
- A hormone precursor (DHEA)
- Lipids, including the typical cholesterol tests and triglycerides
- Liver function using blood measures of albumin, globulin, alkaline phosphatase, and other proteins and enzymes
- Metabolic health using fasting glucose, glycohemoglobin A1c, carbon dioxide, hemoglobin, and triglycerides measurements

We take all these measures together and use the machine learning method we described earlier to estimate a person's overall biological age as well as a biological age for each of their major organs or systems. We chose to use clinical labs because they are well validated and their meaning for health is understood. This maximizes actionability while minimizing cost. It also conveys the delta ages specific for each organ or system that collectively explain the differential health outcomes that, in part, explain why you are biologically older or younger.[23]

This testing can be done for a fraction of the cost of the epigenetic aging test today. Even better, most of the clinical lab tests we employ for the biological age calculation can be covered by insurance as a part of annual physical checkups. As the science of biological aging continues to improve and costs of testing continue to drop, it will be possible to do ever more comprehensive assessments of biological aging—providing more personalized plans and greater power to assess and improve biological aging across systems. We believe this healthy aging opportunity is a striking advance in personal health.

Aging Drugs and Supplements

A person's heart rate is one of the most basic pieces of human health information in the world. It's also highly variable. In adults, a healthy resting heart rate can vary from 40 to 100 beats per minute. That's a huge range, and no piece of data within that range is particularly useful by itself.

The same is true for measures of biological aging. No single measure should ever be overinterpreted; there is plenty of natural variation at play. Taken over time, though, these metrics can help you understand what is happening globally in your body and how you can affect it. As is the case with heart rate, tracking longitudinally—and checking back with your health-action plan—is what makes all the difference.

This is particularly important if you decide to join the growing ranks of people who are taking so-called anti-aging compounds—supplements or low-dose medications that have been demonstrated to extend life spans and health spans across a number of cellular and animal models. What is exciting about studying aging is that the basic control mechanisms and hallmarks of aging appear to be highly conserved across all organisms, from single-celled organisms like yeast to multicelled organisms like worms, fish, mice, dogs, and humans. This conservation means that we can study anti-aging compounds in relatively short-lived animals such as worms, flies, mice, and even dogs and, because these molecules act on biological mechanisms that extend across wide branches of the evolutionary tree of life, it is reasonable to suspect they might work on humans as well. And as more of these compounds pass safety tests, more people are beginning to take them, hoping to facilitate healthy aging. That said, not all supplements are good for you. It is extremely important to clearly understand the evidence behind any claimed supplement or drug and to seek guidance from your healthcare professional.

One promising class of supplements, and certainly the most widely used and researched as of this writing, are NAD boosters—most notably nicotinamide riboside, also known as NR. NR is an alternative form of vitamin B_3 that acts as a precursor for the production of nicotinamide adenine dinucleotide, or NAD^+, a chemical compound that is vital for the process of converting food into energy, repairing damaged DNA, and maintaining circadian rhythms. There is also a lot of current interest in a similar molecule, nicotinamide mononucleotide or NMN.[24] NMN works through similar mechanisms to increase production capacity for NAD^+, and advocates cite that it is one step closer to NAD. That is true inside of a cell and would be fine if it crossed the membrane. However, when taken orally, the body converts NMN to NR to get across the cell membrane, where NR is then converted to NAM and then to NAD^+.[25] NR supplementation has been much more extensively tested in humans than NMN. NR has been shown to boost NAD levels and to be safe; the same has not yet been shown for NMN.[26] (For these reasons, Thorne has focused on NR as its preferred NAD^+ booster.) Because concentrations of NAD^+ decrease in the human body as it ages and low NAD^+ has been implicated in chronic diseases such as heart disease, Alzheimer's, and diabetes, the thought is that boosting the levels of this compound could ameliorate aspects of these diseases and slow the aging process.[27] Indeed, such NAD^+ boosters have been shown to extend health span in model organisms including yeast, worms, and mice.[28]

Another intriguing candidate molecule is rapamycin, which was initially used to treat renal transplant patients as an immunosuppressant, a dampener of the immune systems.[29] Of course we need our immune systems to be appropriately active, so it may be better to think of rapamycin as a compound that, when taken at very low doses, simply eliminates hyperimmunity—keeping our bodies from responding to every foreign insult with a full-on immune attack. Like NAD^+ boosters, rapamycin appears to modulate a master control mechanism for blocking many of the hallmarks of aging, a mechanism conserved across many species, so it may be a promising start for the development of "rapalogs," which mimic its beneficial effects while reducing its potential dangers. Rapamycin has been shown to slow aging and extend life span in model organisms.[30] It is also being used now in patients by a small group of physicians spearheaded by Dr. Alan Green in New York with promising results, but it is not yet widely accepted because appropriate clinical trials have not yet been executed. There are concerns about the immunosuppressive effects of rapamycin, which is why Dr. Green carefully monitors his patients for skin and subcutaneous bacterial infections, which should be treated with antibiotics.[31]

One intriguing finding is that short-term treatment with rapamycin has been shown to rejuvenate the aged oral cavities of elderly mice, including regenerating periodontal bone, reducing bone inflammation, and even shifting the microbiome in the mouth toward a more youthful composition.[32] Another fascinating study by Matt Kaeberlein at the University of Washington showed that giving dogs rapamycin for just ten weeks led to improvement in both diastolic and systolic measures of heart functions, including improved heart ejection fraction.[33] While much still needs to be studied, significant research suggests rapamycin is a highly promising compound for promoting healthy aging. Human clinical trials are ongoing. There are also significant efforts under way to discover rapalogs, including from natural products, that could safely provide these observed benefits over a long period of time.

In the long run, it's likely that a combination of compounds—and perhaps different combinations for different people with different genetic profiles—will be most effective in promoting healthy aging. To that end, a small but intriguing study published in 2019 may be illuminating. In the study, nine participants took a combination of a growth hormone and two diabetes medications for one year. In that time, they shed an average of 2.5 years from their biological ages, as measured through several versions of the epigenetic tests of biological aging.[34] Perhaps any one of those drugs alone might have offered some benefit. Indeed, evidence that they each offered some necessary function to reduce biological aging was why the chemicals were selected for study. But the whole appears to have been greater than the sum of its parts. The study raised its fair share of controversy in the scientific community, however, as growth hormone was earlier touted for its anti-aging potential but was then linked to concerns over increased diabetes and cancers.[35] Metformin, a drug for diabetes, was used as part of the cocktail with the goal of mitigating those risks. Whether the drug combination used in this study can maintain these effects in the long term remains to be seen. Caution is warranted, but the results are intriguing.

Short-term results like these are an intriguing outcome of our emerging ability to measure biological aging with increasing accuracy over shorter times. Previously, to understand the impact of a drug combination on aging, we had to watch for years, even decades. Now we can see an effect over a short time in two steps, a pre- and post-test. Still, the totality of long-term effects, both beneficial and negative, will be unknown for years to come. While some of these compounds have passed clinical trials for short-term safety in humans, we will not know how they affect human life spans and health spans over a long time period for some years, since meaningful verification will require tracking a cohort of people taking these pills over decades. We can, however, point out that, in short-lived animals such as worms, flies, and mice, life can be extended significantly without other obvious side effects. Again, because of the conservation of the aging mechanisms, these results are reassuring for the human outcomes.

Decades, of course, is a long time to wait. It is understandable that many people, having weighed the limited evidence, have decided that these drugs and supplements are worth whatever risks they might bear. But those who decide to wait before embracing these interventions—or who never do embrace them—are certainly not out of luck. Remember, the Arivale participants were able to increase their negative delta ages primarily through lifestyle interventions, guided by their genomes and phenomes. And when we bring together this sort of scientific wellness with attentiveness to long-term brain health, it's not at all unreasonable to believe that 100-year health spans are well within reach of ordinary individuals, independent of aging compounds. But both together may provide additive affects.

And that's going to change things. A book, *The 100-Year Life*, is a fascinating look at what this might mean for us as individuals and a society. In it, the authors, Lynda Gratton and Andrew Scott, describe a world in which many individuals can expect several additional healthy decades of life. What will we do with those extra years? How will we sustain ourselves financially, socially, politically, and creatively? Will we work more? Will social security programs have to change? And how do we ensure that everyone has an equitable opportunity to benefit from increasing life spans and health spans? Society may have to face all these questions—especially the last.

It is important to reiterate that aging is the greatest risk for virtually all chronic diseases such as diabetes and Alzheimer's disease. If we can significantly delay our aging, then we delay the onset of chronic diseases. Thus healthy aging becomes a major factor in preventing chronic diseases and it does so for all chronic diseases at once!

Are we equal to the challenge? We will soon find out. The uncoupling of age and aging doesn't offer us just more years of healthy life but a substantially different idea of what it means to be human.

6

Keeping Our Minds Healthy for Life

Why Neuroplasticity and Cognitive Training Offer Hope for Lifelong Brain Fitness

The world is home to more than 500,000 people over the age of 100, and projections suggest that number will increase about eightfold by 2050.[1] (Even today, people who reach 80 with no clinically demonstrable disease are far more likely than their peers to live another twenty years.)[2] But if we are to embrace all of these extra years of life span productively and creatively, it is essential that we reach old age not only physically functional but also mentally alert. No one wants to spend their last decades healthy in body but feeble in mind, yet our contemporary healthcare system virtually ignores wellness of the brain. Even for those who do not suffer from any form of diagnosable dementia, it is broadly accepted that mental degradation is an inevitable consequence of age.

In time, we'll come to see this as a shameful failure of medicine and human imagination—one of the most dehumanizing presumptions of our modern world. Our brains are, after all, the organs that encode our humanity, control our bodies, allow us to be creative, mediate our interactions with others, and connect us to the outside world. At some level, our brains create the world in which we live. Yet aside from those already suffering from signs of dementia or genetically at risk for such diseases, few people have ever heard their doctors say, "OK, so now let's talk about the health of your brain."

Kidneys and liver? Yes. Lungs? Quite certainly. Heart? Absolutely. Good doctors take many precautions to assess heart health—measuring blood pressure and LDL and HDL cholesterol, taking EKGs, and so forth. We can monitor our heart health throughout adulthood and provide the necessary drugs, diets, supplements, or exercise to help alleviate shortcomings—a tremendous legacy of the pioneering Framingham Heart Study. But everything that could adversely affect these organs—from bad diets and lack of exercise or sleep to smoking and stress—can also negatively impact our brains, sometimes with even more devastating consequences.

Scientific wellness must encompass the health of both the body and the brain. We have been able to replace a person's heart for more than fifty years; we cannot replace their brain. Why, then, do our health systems spend so little time focused on promoting wellness and avoiding diseases of the brain?

First, as with any organ, the brain is a physical tissue, and it exhibits processes associated with health at different levels—but these processes can be awfully hard to measure. Many of the tools researchers previously used, like microelectrode analyses, required opening the skull and using electrical probes to take manual measurements of the brain's operations. Even as noninvasive imaging tools such as two-photon microscopy have come into use over recent decades, the images were seldom precise enough for us to understand what was truly happening in the brain and where it was happening, as fractions of a millimeter can separate neurons wired to do very different things.

You've likely heard the saying "What gets measured gets done." Well, the corollary is also true. What cannot be accurately measured cannot be effectively treated. Until we could take accurate measurements of brain processes (and to be clear, we're still working on that) we couldn't even begin to understand how to keep the brain healthy. We do have powerful imaging techniques today—positron emission tomography (PET) scanning and magnetic resonance imaging (MRI), which permit visualization of metabolic or structural changes in the brain, an important step for brain health. But how often are they used—and *when?*

Our brains are remarkably resilient organs, but that doesn't mean they don't require care. Other organs begin to show signs of disease in our fifties and sixties. Most people remain mentally fit for at least another decade or two after that before their brains begin to suffer from noticeable degradation, but there can be more subtle losses along the way. This wasn't always taken seriously because it was rare relative to other diseases.

The average age today for the onset of dementia is 80. The overwhelming majority of our ancestors never reached that age. It wasn't until about fifty years ago that the average global human life span reached 60. Everybody knew that very old people tended to suffer from cognitive decline, but few people saw addressing the decline as a priority. For most of our history, brain health simply wasn't something we needed to worry much about because our bodies went long before our minds did.

That's no longer the case. The average global life expectancy is approaching 80, and a century of life is no longer a fanciful prospect. But one thing is clear: if we don't keep our brains healthy, longevity will be more of a curse than a boon. Anyone who intends to live a long and healthy life should be determinedly focused on their cognitive health. It matters now more than ever before.

The Role of Plasticity in Lifelong Brain Health

It doesn't take a disease to ruin a good mind. To understand why this is, we need to understand the concepts of plasticity and negative plasticity. Plasticity is the brain's ability to modify its connections in response to changing conditions, reusing the same neurons for different purposes as different needs arise or activating brain stem cells to produce replacement brain cells. At birth, infants have an innate sucking reflex that helps ensure they will be nourished. Eventually, this need goes away, and so does the reflex—most human adults don't make a sucking motion just because something touches their lips. But the neural cells that govern this behavior are long-lasting; some of these cells survive the entire length of a human life. Rather than becoming "dead weight," these cells form new connections with other cells as we take on new information and learn new skills. In addition, the brain has neuron stem cells that can compensate for lost nerve cells.[3] So the brain is plastic and flexible.

In childhood, as the brain works overtime to make sense of the world, the brain's plasticity is engaged full-time. Scientists refer to this neuro-building environment, when new stimuli are continually being processed to generate many new useful neuronal connections, as "blastic." This generally continues into adolescence and through one's late twenties or early thirties.

As we grow older, the habits that permit us to go about our daily lives become more ingrained. Settling into our routines, the blastic nature of our environment stabilizes, and the brain's plasticity is less frequently activated. While it might seem as if the world around us is changing rapidly, most of the moment-by-moment events in our lives are very predictable. We put one leg in front of the other and the ground meets us just as it always has. We walk into a shadow on a hot day, and the world is a little cooler than it was in the direct sunlight.

These are things we learn quite early in life and fuse into our expectations without further thought. In the coming years, more aspects of our lives become rote. Our brains grow used to using doorknobs, tying our shoes, reading a map or graph, or finding the shift key on a computer. Because so much of life becomes increasingly routine as we age, the brain is challenged less. And because so much of life is predictable, the fine details matter less. This process begins in our late twenties or early thirties and grows with each passing decade, becoming ever more noticeable as we age.

This is the beginning of what is known as negative plasticity, a period in which the blastic environment ends and the increasingly predictable nature of life leads to the reduced activation of brain networks and a period of neuro-destructive forces.[4] In this stage, the gradual loss of plasticity means that we can have an increasingly hard time processing new information with the speed and attention of youth, resulting in an inability to separate information that is vital or relevant from all the other information available every second of the day. (And yes, this is very much like the "signal versus noise" challenges we're facing, as we integrate enormous data sets into our personal data clouds.) For this reason, scientists sometimes call this a period of "noisy processing." During this time, humans begin to lose critical synaptic connections—falling into a downward spiral marked by an increasing lack of attention to detail, a narrowing of our perceptive fields, a striking loss of reaction time, and a degradation in most measures of cognitive ability and brain health.[5]

The result is an increasingly rigid fixation on the present and gradual decline of virtually all cognitive functions. Anyone who has ever become irritated after losing their keys—fixated on the fact that "I always leave them right here" rather than immediately beginning the process of retracing their steps—has had a little taste of negative plasticity at work. But while losing keys is annoying, losing cognitive functions—integrated brain activities that can be measured and observed—can be devastating. And that's what negative plasticity does to us over time.

Two of the most important cognitive functions are the ability to attend to sensory input as it comes in (what neuropsychologists are apt to call "attention") and the ability to keep up with information when it comes in quickly (what researchers refer to as "processing speed"). These are the building blocks of all higher cognitive abilities. If you don't attend, you won't remember the information. And if information comes in faster than you can accurately process it, when you try to recall it, you'll find a jumbled mess.

What happens when attention and processing speed are hampered by negative plasticity? We get "old drivers."

There's a reason why insurance rates rise for drivers as they get older. In the first few years of driving, experience sets in, and accident rates begin to fall. But age and good driving do not maintain a linear relationship. The slowing of processing speed means the brain of a driver in their seventies takes in a lot less information in a split-second than a driver in their thirties, narrowing what is called the "useful field of view" so that it's been said that older drivers essentially see the world through soda straws.[6] This helps explain why older drivers' most common accidents occur at intersections, and their most common targets—things like curbs, parked cars, and vehicles in adjoining lanes—are right next to those.[7]

Is all of this inevitable? Scientists long thought so. It seemed like it would be impossible to replicate, let alone maintain, the dynamic, sensory-laden blastic experience of youth. As it turns out, the human brain was desperate to help us identify a solution. All we had to do was listen.

Exercising the Brain

The first artificial cochlear implant was developed in 1957. This was one of the earliest attempts at bionic technology, a year before the word "bionic" was even coined. The need was vast. Even today, hearing loss is the top reason for disability compensation claims among military veterans.[8] In the decade following World War II, it was a virtual epidemic. By the mid-1960s, surgeons had begun inserting cochlear implants on individuals with profound sensorineural hearing loss but, to put it mildly, the products didn't work well.

In the mid-1980s, Michael Merzenich, a University of California San Francisco neuroscientist, decided to focus on the challenge of constructing a more effective artificial cochlea using "sensory maps" to better understand the connections between natural cochlea—which turn sound vibrations into nerve impulses—and the brain. The problem was that there are thousands of such connections. At the time, no conceivable biotechnological creation could have that many attachments, so many experts in the field of hearing assistance had dismissed the possibility of cochlea implants.

Merzenich was not dissuaded. For more than a decade, he had been studying the flexible ways in which the brain responds to new information. He was intrigued by the idea that if surgeons made just a small number of connections in the right places, the brain would work to make sense of the new electrical signals, restructuring itself to make the new pulses decipherable. He thought these advanced implants might restore some sense of hearing in adults over weeks or months or maybe even years. He never expected that he'd be able to have conversations with patients within days of their receiving an implant. This experience profoundly affected Merzenich's thinking about neural plasticity.

Every time a person learns a new skill—from eating food with a spoon to playing a sport to developing the skills of one's trade—their brain changes. At first, these changes are chemical. Eventually, they become structural. Ultimately, they are functional. For a long time, scientists believed that the brain was plastic only in childhood and adolescence, as individuals learned skills and "wired" the relatively blank slate of the brain with which they were born. By adulthood, these scientists believed, the brain was fully developed and "hard-wired." Sure, you might be able to push more information through it—everyone can learn a new skill through hard work and practice—but you could not improve on its operations. In this way, the conventional wisdom held, the adult brain was like a machine—destined to max out and then wear out over time, starting around the mid-thirties.

Who among us over the age of 35 hasn't noticed a gradual increase of temporary mental lapses or failures to apply the correct reasoning to a common situation? Who hasn't become just a little more afflicted by the experience of knowing something but not knowing how to express it—a phenomenon scientists call lethologica (which is closely related to the inability to recall the name of a person, place, or thing, which is called lethonomia)? Who hasn't lost their keys, misplaced their glasses, or forgotten an important date?

Merzenich's revolutionary idea wasn't that none of this was happening—of course it was. But by the mid-2010s, he had come to believe it didn't have to. "At least most (possibly all) plasticity-induced changes are, by their nature, reversible," he and his collaborators wrote in 2014 in an article in *Frontiers in Human Neuroscience*, two years before he would go on to win neuroscience's highest honor, the Kavli Prize, for his studies of brain plasticity.[9]

Since 1949, when influential neuropsychologist Donald Hebb suggested that neurons were adaptable over time—a rule often summarized as "neurons that fire together wire together"—brain researchers have found evidence that supports the notion that neurons that are repeatedly active at the same time will tend to become "associated." Because of this phenomenon, Merzenich wrote, "it is just as easy to degrade the brain's processing abilities as it is to strengthen or refine them. In the designs of therapeutic training regimes, the Hebbian 'rule' must be considered, to assure that training-driven changes are always in the positive, strengthening, recovering, re-normalizing direction."

Even for someone with a history of making big claims, this was a bold assertion. Sure, with the help of a complex piece of invasive technology, the human brain could figure out how to interpret a limited set of signals to reacquire a primary human sense, hearing, but Merzenich was making a much broader claim for the power of plasticity. What's more, the claim was largely made on the strength of experiments involving rodents, whose brains are very different from our own. (Suffice it to say, our species has moved on quite a bit past mice.) But the results of these rodent studies were compelling.

"The rat listened to a series of frequencies, and every time the rat heard the target frequency, the rat could get a food reward," Merzenich and collaborator Karlene Ball explained in 2017. "As the rat got better and better at noticing the target sound amidst distractor sounds, we made the sounds faster, and less different—to improve auditory precision and speed."[10] Using this model of persistent "strengthening," "recovering," and "re-normalizing," the researchers managed to restore youthful brain activity in older rats—something that shouldn't have been possible under previous assumptions of non-plasticity for mammalian brains.

Exercises that continuously and progressively challenged the speed and accuracy of brain processing didn't just result in improvements in the rodents' ability to complete the task. Advances in brain imaging allowed the researchers to see changes in neural wiring, neural coordination, and neural precision. In fact, every aspect of a rodent's brain health that the researchers could measure—some twenty different cognitive characteristics in all—had been changed by the brain training.

The study results bolstered Merzenich's long-held belief that age-related losses in human brain plasticity, as measured by a few dozen easily quantified cognitive functions, could be restored to a more youthful functionality through intensive, repetitive, and increasingly challenging training.[11] While trigger sounds and food rewards might not be sufficient for humans, Merzenich hypothesized that an "adaptive cognitive training" program based around well-designed computer games could do the trick, and that it would outperform other types of casual brain games.

Merzenich and his collaborators described the results of a study to investigate this hypothesis.[12] In the experiment, adults with an average age of 70 were randomly assigned to spend forty-two minutes a day on computer games. One group played casual games of the sort many people have downloaded on their phones to while away the time. The other group engaged in games specifically built to provide adaptive cognitive training, analogous to progressive overload exercising for muscle growth. Importantly, the people in both groups were "cognitively normal" adults. Ten weeks later, study participants completed tests that measured their processing speed, working memory, and executive control. Those who were assigned to the cognitive training group far outperformed those who had just been playing the "fun" games. In short, the treatment drove significant improvements in neurological function.[13]

Consider what significantly improving cognitive function would mean for people in their eighties who we might think of as very healthy "for their age." What would it mean to be able to combine eighty years of accumulated wisdom with the lucidity and the integrative powers of thought and responsiveness of a much younger mind?

It would be game-changing.

The Benefit of "Brain Games"

Traditional cognitive maintenance has historically been limited to tips and strategies intended to help people compensate for the gradual decline and seemingly inevitable loss of flexible brain functionality. "Do crossword puzzles," a doctor might say. "Vary your route to and from work every day," another might advise. "Brush your teeth with your non-dominant hand." This is apple-a-day advice. There's nothing wrong with it. It might even be helpful to some extent. But it's no substitute for a truly blastic environment.

The building and maintenance of muscle is a fitting analogy. Many of us remember with wistfulness the days in our teens and twenties when maintaining a healthy physique was simple. The very fabric of our lives—walking to school, playing on sports teams, exploring the world on a bicycle—naturally helped us stay fit. A youthful metabolism didn't hurt, either. As the decades come and go, however, few of us can remain fit without some degree—often a considerable degree—of intentionality. We run. We go to the gym. We attend yoga classes. We diet and lapse and diet again. It can be a lot of work to maintain healthy habits.

The exciting thing about the emerging research related to "brain games" is that, while adaptive cognitive training is more work than doing an occasional crossword puzzle, it's not an altogether different experience. Designed right, it can be both beneficial and enjoyable. This is like finding a diet and exercise regimen you actually enjoy—it makes the effort of staying fit a whole lot easier.

One example is an exercise from a brain-training subscription service, BrainHQ, that is part of a company that Merzenich helped launch in 2005, the Posit Science Corporation. In the game, called *Hawk Eye*, a group of birds flashes on the screen across a blue sky. One bird is different; that's the target. The others are all the same; those are the distractors. After the flash of the image, the player must identify the location of the target bird. In successive viewings, the birds change location. As the player improves in accuracy, the viewing window shortens, measured in milliseconds. At higher levels of play, the shape, colors, tint, and reflectiveness of the birds vary and become increasingly similar, and the background becomes more complex, with lower contrast and varying lighting, appearing more and more like the real world.

Does that sound a little like an upgraded version of the 1980s Nintendo game *Duck Hunt?* Well, there is no laughing dog, but it's true that the experience isn't all that different. What is different is the result. While not all computer games are equally adept at fighting negative plasticity, ones designed specifically for that purpose can be very effective. Training in this visual processing ability has been shown to predict fewer traffic accidents in the elderly and to reduce risk of dementia.[14]

Can't life itself stimulate the rebuilding of neural connections? Can't we simply make life more blastic? Sure, to some extent. It's great if you enjoy taking a leisurely walk around your block in the morning and eating salad every day. But those things, by themselves, aren't enough to keep you mentally fit. Similarly, playing *Call of Duty*, *The Legend of Zelda*, or *Tetris* on your computer probably isn't going to be enough to keep your brain fit. Commercial video games may have some benefits, but they aren't designed to address a health need. Brain training games like *Hawk Eye* are. These games are created with the specific goal of addressing plasticity—training sensory perception and focusing on things such as speed and accuracy of processing. They help players identify the limits of their capacity and increase it, bit by bit, using progressive overload. That is, they find the very limits of what you are capable of for a particular task and push you to that limit—much like weight training pushes your limits to achieve muscle growth. This has tremendous promise to impact a wide array of everyday functions.

Yes, we're still in the early days of understanding what works best to improve the chemistry and structure of the brain. And it is important to note that there has been significant debate about the degree to which users of these games are getting better at the specific task versus generally increasing cognitive performance overall.[15] But that's another exciting part of this revolution: early attempts to build adaptive cognitive training programs in the form of computer games have already seen quite a bit of success.[16] To date, Merzenich's approach has been tested extensively in over 100 published papers.[17] What comes next is likely to be even more beneficial.

When he first conceived of the exercises that became the starting place for the development of BrainHQ, Merzenich was hoping to help reverse the typical cognitive decline of people in their eighties. It wasn't long before researchers recognized that the exercises delivered about the same degree of improvement to people in their seventies and sixties. And given that we now know that cognitive decline begins for most people decades before then, it shouldn't come as a surprise that many of the service's subscribers are in their thirties and forties.

You've probably heard of at least one of them.

There was just 1:30 left on the clock in Super Bowl XXXVI when Tom Brady, then 24, took the ball to begin a drive that would set the New England Patriots up for a last-second field goal and a 20–17 victory over the St. Louis Rams. It was Brady's first Super Bowl victory. He earned another two years later. And then another a year after that. The Patriots made it to the Super Bowl again in 2008 and 2012, losing both times. By 2014, many football pundits had come to believe Brady's best years were behind him.

Brady didn't agree. He believed that, with the right physical conditioning, recovery, and nutrition, he could not only keep playing but continue to improve. As he entered his mid-thirties, he was of the mindset that keeping his body healthy wasn't enough—he needed to do something to sharpen his mind.

Posit Science chairman Jeff Zimman will never forget the day he got a phone call from Brady's longtime health coach, Alex Guerrero, shortly after the 2014 Super Bowl. Many pundits agreed it was a "down year" for Brady, who had struggled in early-season games. Zimman learned that Brady had begun using BrainHQ in an attempt to sharpen his mind and believed it was improving his on-field performance.[18] Soon, Zimman and Posit's CEO, neuroscientist Henry Mahncke, were on a plane to Boston, where they met Guerrero and Brady at the TB12 Sports Therapy Center. Zimman visited the Patriots about every six weeks over the next year and had many conversations with Brady about his comprehensive health regimen. Brady told him he was in better physical condition at 36 than he had been ten years earlier. What about cognitive conditioning, Zimman wanted to know.

"I'm in the best cognitive condition of my life," Zimman recalled being told. "I can see more. I can see it faster. I can make faster decisions."

Talk is cheap. On the gridiron, only results matter. Sure enough, less than a year later, Brady was back in the Super Bowl, where he led the Patriots to a 28–24 victory over the Seahawks (much to our chagrin as Seattleites). Two years later, he was back again, leading his team to a 34–28 overtime win over the Atlanta Falcons and capping a season that many believed was his greatest ever. He returned to the Super Bowl the following year, when his team lost to the Philadelphia Eagles, and then again the year after that, when the Patriots defeated the Los Angeles Rams 13–3. Brady now has seven Super Bowl victories—far more than any other professional football quarterback.

BrainHQ exercises have been shown to improve performance across a wide range of workforces.[19] One study showed improvement in the ability of law enforcement officers to distinguish scenarios in which they should shoot from those when they should hold their fire.[20] Another looked at ways of enhancing attentiveness among electric power line workers.[21] Researchers noted gains in productivity, accuracy, inhibition control, safety, and overall cognitive efficiency.

BrainHQ isn't the only commercially available training program. In 2017, a team of Australian researchers identified eighteen such programs, seven of which had been evaluated in a research setting and which exhibited at least some effectiveness in improving cognitive fitness. BrainHQ was one of two programs that met "Level 1" evidence, based on properly designed randomized controlled trials. The other was Cognifit, founded by psychologist Shlomo Breznitz. Cogmed, Brain Age 2, and My Brain Trainer met "Level 2" thresholds, but this doesn't necessarily mean these programs didn't work as well—it simply indicates that more systematic testing is needed.[22]

What we can say with great confidence is that you don't have to be a seemingly superhuman football player to have a brain that stays fit with age. You also don't need to use a computer program to enjoy the benefits of renewed plasticity. People who take on a new regimen of exercising or dancing have been shown to enjoy increased plasticity.[23] The same appears to be true for those who learn a foreign language later in life.[24] Research also suggests that individuals who have had musical training throughout their lives enjoy greater neuroplasticity in their elder years, and some studies are now under way to see if musical training that begins later in life will have a similar effect.[25]

What Our Brains Need

Brain exercise, like physical exercise, is good for just about everyone at any point in life. In much the same way that we can now identify what kinds of exercise different individuals need based on their genome, phenome, and specific lifestyles, it is likely that we'll soon be able to identify the specific brain training regimens individuals need either to stay cognitively fit or to regain cognitive fitness when it has been lost. Slowly at first, and then quite rapidly in coming years, the combined power of individual data clouds and artificial intelligence will lead us to a world in which a combination of a cognitive assessment, whole-genome sequencing, analysis of a person's blood analyte composition and gut microbiome, and various digital health measurements will produce an individualized road map for neurological (as well as physical) fitness.[26] The gut-brain axis in particular is emerging as a fascinating regulator of brain health, where the microbial community in the gut can actually impact neurochemistry, cognition, and behavior.[27] We will be learning how to harness this powerful tool in the coming years. Just as drivers have come to expect smartphone-based mapping apps to readjust when we make a wrong turn, real-time data analysis will readjust our brain training interventions based on day-to-day assessments of cognition.

We know that even the most basic tools can be used to address cognitive issues across a wide range of individuals for a wide array of conditions. First and foremost of these is nutrition. When researchers from the Netherlands assessed the diets and physiological brain health of nearly 4,500 people, they found a striking correlation: those who eat lots of vegetables, fruit, whole grains, nuts, and fish and avoid sugar-containing beverages had larger brains, more gray and white matter volume, and healthier hippocampi (a part of the brain that is involved in forming, storing, and processing memory).[28] There are also nutrients that can significantly reduce cognitive decline in individuals suffering from concussions, some of which Thorne co-developed with Mayo Clinic neurologist David Dodick into a product called SynaQuell, which was validated for improving outcomes for concussed athletes against a control group in a Mayo Clinic-led clinical trial.[29]

"There's this kind of neglected component of food that is really the effect that our nutrition diets have on the way our brains work," noted

Dr. Lisa Mosconi, author of *Brain Food: The Surprising Science of Eating for Cognitive Power* in a 2017 interview. "Food is not just the source of nourishment and entertainment, but food is really chemistry."[30]

One example that has found its way into the popular press—for good reason—is omega-3 fatty acids. Omega-3s are incredibly important for brain development and continued brain function throughout life, and some forms are more highly concentrated in the brain than any other tissue in the body. Omega-3 deficiencies over long periods are associated with cognitive decline and mood disorders, including major depression. Omega-3s are found in many healthy food sources, including fatty fish (which actually get them from eating marine algae).

Another example is the B-vitamin known as choline, which is found in many foods including chicken, fish, beans, broccoli, and peas. Choline is a precursor for acetylcholine, an essential neurotransmitter involved in circadian rhythm, memory, and other brain functions. Another form, phosphatidylcholine, may aid in delaying dementia (more on that later) as long as you don't have certain known bacteria in your gut that convert it ultimately to a harmful substance known as TMAO.[31] But choline isn't unique—nearly every food we eat has an impact on our body's chemistry, and some in turn have an impact on our brains.

After diet comes exercise—and we've known for quite a long time how important exercise is for brain health. A research review from the University of California Irvine's Institute for Brain Aging and Dementia left little doubt about how important active lifestyles are to healthy brains. "Exercise sets into motion an interactive cascade of growth factor signaling that has the net effect of stimulating plasticity," the researchers wrote, noting that exercise also enhances cognitive function, reduces many of the causes of depression, stimulates the creation of new neurons, and improves healthy blood flow in the brain.[32] (We will dive much deeper into new understandings of the mechanisms of Alzheimer's disease in Chapter 8 to make clearer why exercise is so protective of the health of our brains.)

Our brains also need sleep, and the research is very clear regarding what happens when we are sleep deprived. Whether it is a chronic problem or a more occasional issue, sleep deprivation impairs attention, impacts short- and long-term memory, affects decision making, reduces vigilance, and affects brain health.[33]

Finally, our brains need interaction with others. Nobel Peace Prize–winning cleric Desmond Tutu often pointed to the Bantu word "ubuntu" to explain how important community is to individual personhood. The word roughly suggests that "a person becomes a person only through other people." Tutu, who passed away in 2021 at the age of 90, noted that a person with ubuntu knows that "he or she belongs in a greater whole." Since our brains are the organs through which we come to understand ourselves, it might be said that our brains rely on the greater whole. This notion is backed by plenty of research, including a study from an international team of researchers showing that cognitively normal Chinese elders who engaged in a regimen of social interaction and exercise showed significant increases in total brain volume as well as improvements in several neuropsychological measures.[34]

Nutrition, exercise, and social engagement also impact another key area of brain health intimately related to cognition: mood health, especially depression and anxiety. These disorders, extremely common today, impact cognitive function in the moment and carry the risk of cognitive decline in later years. Emotion is a major factor in neuronal plasticity. Events that are linked to strong emotion are deemed important by the brain and much more likely to be encoded as memories. Depression significantly dampens the "plastic" environment, increasing the rate of cognitive decline. Anxiety has a similar effect on brain plasticity by elevating stress hormones such as cortisol that actively work against plasticity, especially in the hippocampus.[35] For a deep and accessible view into this fascinating topic, we recommend *Why Zebras Don't Get Ulcers* by Stanford professor Robert Sapolsky.

A key component to maintaining lifelong brain health is keeping a close eye on mental health and seeking treatment when necessary. There are many approaches to mood health, from formal medical assessment for major depression to dozens of phone apps to track anxiety and other symptoms of lesser dysregulation. All the health behaviors discussed above contribute to mood health for many of the same reasons they contribute to cognitive health, as these two domains of brain function are largely inseparable.

If a hundred years of healthy life appeals to you, much of the power is in your hands. Cognitive decline and mental health disorders are not inevitable, and our future healthcare must embrace this fact. But to achieve such long-lived cognitive vitality, we will have to do more to stop the degenerative brain diseases that ravage our brains as we age. If the idea of overcoming Alzheimer's seems impossible, that's understandable. Humankind has thrown billions of dollars into researching treatments and therapies to stop this terrible disease—and almost nothing we've done to treat it so far has been worth a damn.

But that may be about to change.

7

The Long Goodbye

How a Personal Tragedy Drove a Dedication to End Alzheimer's Disease

Nathan and I share a vision for the future of health, which is why we have been sharing a voice throughout most of this book. But this is a chapter I need to write alone, for it is a very personal story about a long goodbye—one that my children and I have been saying, again and again, for more than seventeen years.

My wife, Valerie Logan, was small, vivacious, kind, and generous. She shared a lifetime of mountaineering and rock climbing with me, and she was "tough." She was an engaged and committed mother with enormous energy and drive. We both loved traveling across the world with our kids. When we moved to Seattle in 1992, Valerie took responsibility for the K–12 education programs of the new Department of Molecular Biotechnology at the University of Washington (UW). When I left the UW to form the Institute for Systems Biology, she took on the same role at ISB. She helped secure a series of local grants from the National Science Foundation that fueled our future efforts. Her work catalyzed fundamental changes in science education, first in Seattle and then across the state of Washington. ISB's education center was later named the Logan Education Center in honor of her leadership and accomplishments.

Valerie was a force. She could have done anything in the world. The fact that she wanted to live her life with me was one of the central pillars of my happy adult life.

Around 2004, I began to notice that Valerie was starting to forget things. She was 67, and I figured it was simply a consequence of getting older. We were aging, after all; there was no denying that fact. Few things make you feel older than watching your children get married, as our son and daughter had recently done. Valerie had been pushing for a move from our beautiful lakeside home, which was becoming too much for her to manage. She had a lot on her plate, so it seemed perfectly understandable that she was forgetting things here and there. I assumed I probably was, too, and just didn't realize it.

There's no need to be coy about what happened next. Even if you didn't read the title of this chapter, you'd know by now—for Alzheimer's isn't some strange and rare disease, but a fact of life for tens of millions of people and their families around the world. You've likely heard a story like this before. You may even be living through it right now.

Valerie's diagnosis came in March 2005. The news was devastating. There really is no other word, and even that doesn't do the experience justice. I'd been dragging my feet on the move; I loved our home on Lake Washington. But after Valerie's diagnosis, the decision was obvious: we moved to a condominium in downtown Seattle where minimal housekeeping was necessary. This was when it really became clear to me how much Valerie had already been affected by her disease. In the past, she would have managed the organizational aspects of such a move seemingly without effort. Now, she was struggling, confused, and vulnerable in a way I'd never seen in the forty years we'd been married. On most days, she was fully aware of the ways in which her mind was beginning to slide, and my heart ached for her.

She did manage, over the next few years, to hide her disease from outsiders. It was truly impressive. But in the long run, her natural intelligence was no match for this insidious, neuron-destroying disease. She was no longer able to read as fast as she had, so she opted out of our book club. She stopped playing games with old friends. Perhaps the biggest crisis at that point in our marriage came in 2009, when I had to take her car keys away after a series of fender benders. It was unquestionably the right thing to do, but it came with an enormous loss of freedom. She still went out on walks, but it wasn't long before she was getting lost. Soon, it was no longer safe for her to leave home alone. Our condo concierges became adept at intercepting and returning her when they saw she was trying to leave the building. Our home was becoming her prison.

As often happens to people who suffer from Alzheimer's or other forms of dementia, her personality began to change in ways that were hard for those around her to accept. She became short with me, and angry, and her outbursts extended to our children and grandchildren. In her clearer moments, she was still the person I'd known since high school, but those moments were growing further and further apart.

The Struggles of Caregivers

Once, when I went to a meeting for Alzheimer's caretakers, I sat in a circle with eleven others and listened to their stories. "I have no life apart from caretaking," one said. "I am not going to have any resources left for retirement," another added. And then, again and again, the saddest refrain of all: "I just want this to be over."

I'd always recognized how fortunate I was to have the resources to manage Valerie's Alzheimer's journey as gracefully as possible. On that day, however, I began to feel a sort of survivor's guilt that has never completely left me. The partners of individuals with Alzheimer's, and especially those with early-onset Alzheimer's, are often forced to choose between working and staying home to care for their loved one. Either way, the financial strain can be ruinous.

I was able to keep working, a privilege that so many in my position do not share, in large part because we could afford to hire people to care for Valerie. One of these amazing caregivers, Joanne Fiorito, became nothing less than a member of our family from the time she joined us in 2010. She and Valerie would go on long walks. They would sing and dance together. They talked and ate and read books. Even as the changes Valerie was experiencing were putting a strain on our relationship, Jo was always able to see the best in her. "Valerie has a beautiful spirit," Jo once told me. "She is caring and intelligent and enthusiastic—and I am so happy I've gotten to be a part of her life." Watching Jo come to love Valerie helped me remember all of the reasons I so deeply loved my wife.

By 2015, it was clear that Valerie needed round-the-clock care. After a careful search of memory homes, we moved her into a center about twenty minutes from our condo. Joanne continued to visit Valerie on a daily basis and to take her to lunch and on walks. Valerie made two good friends at the home—and the music and educational environment was very rich. But her neurological health continued to decline, and she soon stopped speaking. Eventually, it was no longer clear if she recognized anyone who came to visit, apart from Joanne. I found it incredibly difficult to watch the woman with whom I had shared my life drift slowly away and know there was nothing I could do to stop it.

In January of 2019, Valerie fell and fractured a lumbar vertebrae, at which point we moved her to a smaller dementia home where she could get even more intensive care. She never really recovered, and has not walked since. As I write these words, Valerie is still with us and now in hospice care. Joanne and I still visit her on a regular basis. But I do know what is coming. I already know how much I will miss her when she is gone, as I have been missing her for these many years.

A Legacy of Failure

The first question that entered my mind after learning of Valerie's diagnosis—the one I posed to my medical colleagues—was "What therapies will be best?" The answers were tremendously disappointing. There were some standard drugs that might help a little initially, I was told, but nothing could stop it. Nothing.

This wasn't only discouraging to me as the partner of someone with Alzheimer's disease. It was disheartening to me as a scientist and maddening to me as a taxpayer—for there is perhaps no clearer example of the extreme costliness and rampant failure of twentieth-century medicine than our collective inability to help those with Alzheimer's.

Since the early 2000s, there have been more than 500 clinical trials for Alzheimer's drugs. Most of these trials have been focused on the most apparent aspects of the disease pathology—amyloid plaques and neurofibrillary tangles, which are composed primarily of beta-amyloid and tau, respectively. Address these issues, the logic went, and you could slow or stop Alzheimer's. Yet every one of these 400 trials has failed.

The United States Congress approved nearly $3 billion in funding for the National Institutes of Health to study Alzheimer's disease in 2020. Since we spend more than $300 billion dollars a year caring for people with Alzheimer's and other dementias—and their caretakers rack up another $250 billion each year in lost income—that could be money well invested. But what good does it do to invest that sort of money on the same old failed approaches?

It's true that finding effective treatments for a disease is extremely difficult. Is it possible the tides could turn if only we do more trials? Many more are under way, so we'll soon see. If a drug company were to discover a medicine that actually worked, it would change the world. Lest we fool ourselves, we should acknowledge that pharmaceutical companies are not solely motivated by the opportunity to save lives: they are businesses, and the financial rewards of developing an effective treatment for Alzheimer's would be tremendous. If a cure is out there, it would be worth several fortunes. But in what is perhaps a sign of growing hopelessness, drug companies that routinely make long-odds bets in hopes of big rewards are pulling back from Alzheimer's research. Pfizer, one of the largest pharmaceutical companies in the world (and the developer of one of the most effective COVID-19 vaccines in record time) announced in 2018 that it had completely dropped out of the race to develop anti-neurodegeneration drugs and has been selling off its stock of candidate Alzheimer's drugs to other companies for a fraction of what it spent developing them.[1] Many other companies quickly followed suit.

Is Alzheimer's simply a disease that can never be effectively treated? I do not believe so. Not for a moment. But we've been looking at this problem in the wrong way. We have, by and large, been looking for a drug, or a combination of drugs, that will treat this disease once it arises.

By now you will have a better grasp of why the research enterprise that supports our healthcare system works the way it does. You will understand why researchers are looking for the "magic bullet." And you will understand why it makes sense, in a twentieth-century sort of way, that almost all Alzheimer's research efforts have been focused on people who are already highly symptomatic. But when you look at people who are in the late stages of a disease—or the tissues of those who have already succumbed to that disease—it's very hard to see anything aside from the multitude of reflections of the illness. It's all symptoms rather than cause. So it is that beta-amyloid and tau look like the cause, when in reality, they might simply be part of a long chain of events that have little to do with how the disease began.

Perhaps we should not completely give up hope that there could be a drug out there that might stop or slow Alzheimer's, but hopes are dwindling when it comes to the "amyloid hypothesis."[2] Some scientists believe there is promise in immunotherapies that target tau, which is more clearly correlated with synaptic dysfunction, but there has been a fair share of disappointment down that road as well. While any advance would be very welcome, it seems highly unlikely that Alzheimer's will be solved by a therapy focused on a single drug target.

Beyond the Search for Drugs

So where does one go next in this quest for an Alzheimer's cure? The challenge is daunting. Alzheimer's is highly complex and arises from multiple disease-perturbed processes and causes. Its symptoms present and progress differently in different people. But a number of important studies are under way that are generating dense molecular, physiological, and clinical data at a scale that dwarfs what was previously possible. This includes large-scale efforts sponsored by the National Institute on Aging such as the Accelerating Medicines Partnership Program for Alzheimer's Disease, for which Nathan has served as a principal investigator. These are generating and analyzing massive amounts of high-dimensional data sets. So instead of looking at one or two potential causes such as beta-amyloid and tau (which probably are not even causal) researchers are increasingly focused on the complex biological networks that are perturbed. These researchers follow each perturbation upstream to places where, with luck and determination, we may identify early interventions that can prevent Alzheimer's from taking over.

A longtime leader in the field of neuroscience, Dale Bredesen, worked in a lab at Caltech as an undergraduate while I was there. After receiving his MD, he decided to focus on Alzheimer's and became the founding president and CEO of the Buck Institute for Research on Aging. When we ran into one another at a conference in 2014, shortly before Valerie moved to the memory home, I told him about my personal interest in Alzheimer's and expressed my frustration with the limited range of options. What he told me changed my view of what might be possible with this disease.

Bredesen was focused on developing a systems approach to reversing the loss of synaptic communication between nerve cells that is central to the mental degeneration associated with Alzheimer's. From this analysis, he categorized thirty-six factors to help improve synaptic communication and suggested that bringing these to normal levels in Alzheimer's patients might be the beginning of an effective treatment. These factors included appropriate diet, reasonable exercise, adequate sleep, supplements where necessary, stress reduction, balancing or addressing clinical chemistries in the blood, and eliminating environmental toxins like black mold. Note that these solutions aren't "drugs"—they include modifying lifestyle features, normalizing blood analytes, and eliminating environmental toxins, an approach very similar to scientific wellness. Bredesen also described different subtypes of Alzheimer's, each of which led to different combinations of treatments. This point of view was eye-opening for me, for while I had been told by many others that there were many different symptoms and many different treatment options for Valerie—none much more promising than the next—no one had suggested to me that she might have just one of the many types of this disease and to correct the disease, once acquired, many approaches would be required.

Each patient with Alzheimer's disease, as Bredesen illustrated it, is like a barn with many potential holes (deficiencies) in the roof. Different patients will have different combinations of these holes, so it does little good to patch just one or a few of them, or to follow the same course of treatment for every patient. A cure will only come if we can figure out how to identify the holes for each individual patient and then plug the right holes.

Bredesen came to believe that biochemical and physiological analyses could help identify many of each patient's particular set of holes, thus offering a map for appropriate combination therapies—what is known as a *multimodal* approach. His protocol, which he called RECODE, varies from patient to patient, but it includes a combination of diet, exercise, correcting biochemical blood deficiencies, improved sleep practices, stress reduction, brain stimulation, and oral care (because research suggests that a healthy oral microbiome can help minimize one of the most common routes pathogens take to get to the brain.)[3] He claimed that hundreds of early-stage Alzheimer's patients who were placed on this protocol had markedly improved; his experiences are described in his 2017 book *The End of Alzheimer's: The First Program to Prevent and Reverse Cognitive Decline.*

Bredesen's claims have been controversial, to say the least. University of California San Francisco neurologist Joanna Hellmuth is one of many researchers who has taken issue with his research methodology and the bold claims made in his book. "Hope is important in the face of incurable diseases and intuitive interventions can be compelling," Hellmuth wrote in her *Lancet Neurology* review of Breseden's book, a *New York Times* best-seller that remains popular many years after it was first published. "However, unsupported interventions are not medically, ethically, or financially benign, particularly when other parties might stand to gain."[4]

It was a harsh but understandable critique. Bredesen's initial published report was based on a nonconsecutive series of ten patients—where the positive results were reported but not the negative ones, meaning there was no way to assess the effectiveness of the program.[5] These reports were discounted by most Alzheimer's experts. Bredesen has written that this disapproval initially complicated his path to funding for clinical trials. Ultimately, he was able to gain support for a trial that used each patient's genome and phenome to guide the selection of a personalized protocol. Of twenty-five patients with early-stage Alzheimer's disease or mild cognitive impairment, the researchers reported in 2022 that twenty-one improved, fourteen markedly so, during the initial study period.[6]

That small study is a start, but it doesn't necessarily vindicate Bredesen's approach. There's much work that still needs to be done if his methods are to be validated with large-scale, impartial clinical trials. In 2021, he followed up with a peer-reviewed report of a larger trial of 255 people using his RECODE protocol.[7] The effect wasn't as high as in the proof of concept study with twenty-five people, but the researchers were able to see a modest improvement in cognition for those in the earliest phases of the disease. Such a result compares more than favorably to any Alzheimer-specific drug treatment available now.

Bredesen hurt his case by failing to disclose results of patients who had shown no improvement in his initial study, but I still feel that his work has been influential and that it points in a direction we need to pursue further. What he has offered is a path toward a new kind of therapy—a personalized, systems-driven, lifestyle-informed, multimodal approach to dealing with early Alzheimer's. This coincides with my own view that chronic disease needs to be managed with personalized multimodal therapies, applied early in the disease process—an approach on which I have come to believe the entire field of medicine, and especially treatment of chronic diseases, should be built. Such a multimodal approach is being further validated by far more detailed systems and computational analyses of Alzheimer's disease and its patients, as we will show in the next chapter.

The key factor as to why I believe we can make a big difference in preventing Alzheimer's and treating it at its earliest stages lies in the results of the only large-scale randomized trial that has really made a significant difference. To date, the strongest evidence supporting a multimodal approach to Alzheimer's disease and other dementias comes from a team of Finnish and Swedish researchers who launched the Finnish Geriatric Intervention Study to Prevent Cognitive Impairment and Disability—the FINGER trial.[8] This two-year, double-blind randomized controlled trial included 1,260 people experiencing symptoms of cognitive decline but were not yet diagnosed with clinical impairment. They were placed into a multimodal therapy program that included dietary guidance, exercise, cognitive training, and social activities, along with intensive monitoring and management of metabolic and vascular health. In essence, the participants who were selected to receive the interventions underwent something akin to a scientific wellness approach like the one we facilitated in the Pioneer 100 study, which was happening at about the same time (albeit without the big data component).

There was some skepticism when the FINGER study was announced in 2015. Many of the specific interventions had been tried before on dementia patients, often with disappointing results. But study coordinator Tiia Ngandu and her team weren't convinced that these approaches were all dead ends. Exercise alone might not move the needle enough. Cognitive training by itself might have mild effects. Socialization might not do much to halt the terrible march of Alzheimer's. But Ngandu reasoned that the whole might be greater than the sum of its parts. *And it was.* By the end of the trial, the patients who had received the multimodal interventions were far more likely to have maintained—and even improved—their cognitive functioning. Their memories were stronger. Their ability to pay attention and maintain that attention was better. The relationship between their thoughts and physical movements was sharper. These were people who were already well down the path to dementia—for whom the wellness-to-disease transition was many years deep—but who hadn't yet been diagnosed with dementia.

Not only does the FINGER trial provide evidence for the efficacy of a multimodal approach to reducing dementia, it was the most successful trial ever for Alzheimer's disease, far outperforming all previous drug trials. "Now we have evidence from the FINGER study showing that, in a group of older adults who are at risk, multimodal interventions have a beneficial effect on cognition," Ngandu said a year after the study was completed. "That's, of course, in an older population. But does it work on memory clinic patients? We have to move forward to see. We need more study on those populations."[9] The attractive idea that emerges from these studies is that we can use multimodal therapies in patients at high risk for Alzheimer's disease (but still functioning normally) to prevent disease transition.

You might correctly have inferred from Ngandu's words that there is an important question at play when it comes to multimodal therapies: How late is too late? With more research, we may soon know the answer to that question. We might also learn answers to the question's corollary: When should these sorts of interventions begin?

Understanding Our Risk

As the life partner of someone who is a shell of the woman I fell in love with, I viscerally understand the desire to help those who are suffering. If we suddenly found a drug that could reverse Alzheimer's at its late stages, I would rejoice. Yet I feel strongly that we should also be working at the other end of the progression—doing everything we can to identify people who are at risk of Alzheimer's or in the very earliest stages of the wellness-to-disease transition, long before clinical symptoms begin to emerge. We need to figure out how to hold degradation at bay or tip things back to full health.

The fact is that the greatest potential for slowing and stopping any disease is in reversing the disease in people who are still in a state of wellness or in the earliest stages of transition—years, likely decades, before the arrival of far more obvious clinical symptoms. This is precisely what we are doing with the integrated approach to Alzheimer's that Nathan and I will describe together in the next chapter. It is clear that a technique employing metabolic positron emission tomography (or PET scan) can detect changes in the brains of potential Alzheimer's patients ten to fifteen years before the disease is clinically detectable. PET scanning is expensive and employs potentially dangerous radioactive exposure, and we hope to be able to find simple blood proteins that mark this earliest transition. When we find these blood protein biomarkers, we will be able to identify patients who have transitioned to the earliest stage of Alzheimer's and use multimodal therapy to reverse the disease at its earliest transition stage.

These efforts will be a model for dealing with many other types of chronic disease, including obesity, cancer, heart disease, lung disease, autoimmune disorders, and diabetes. But just as is the case with those other diseases, while we wait for this "early prevention" revolution in medicine to take hold, there are many things we can do right now.

Given what happened to Valerie, and the fact that we have a history of dementia on both sides of our family, you can understand why our children, Eran and Marqui, would want to know if they carried the Alzheimer's risk gene. Marqui was especially concerned, and rightfully so, as Alzheimer's affects women more frequently than men, and it had affected both her grandmother and her mother.

All of our family members have been tested for the APOE4 variant that is the strongest genetic risk factor for Alzheimer's. Valerie and I both have one copy of the bad gene. That increased the odds that our children would inherit one or two copies, although it doesn't make this inheritance a certainty. To Marqui's great relief, she did not inherit a copy, and this knowledge lifted a heavy weight from her shoulders. This doesn't mean that she will never get Alzheimer's, but her risk isn't elevated by the APOE4 variant, as her mother's was.

Eran, however, has two bad copies. This puts him in a group of extreme risk. To be certain, this is a great burden to carry. But it also offers him a tremendous incentive to do everything he can, while he is still relatively young (53) to prevent the transition from occurring. It is now well understood that our genes are not our destiny. How we live our lives can have a tremendous impact on whether a genetic predisposition for a disease ever materializes.

Fortunately, Eran has always been in tremendously good shape physically. He was running ultramarathons, hunting deer and elk, and rock climbing throughout the West long before he learned that exercise confers significant protection against Alzheimer's. In the wake of his test, he began working with one of Bredesen's collaborators, Mary Kay Ross at the Brain Health & Research Institute in Seattle, to build a personal multimodal program aimed at prevention. None of this guarantees he won't eventually get this disease, just as putting on a seatbelt doesn't guarantee protection against a deadly car accident. But we have a choice every time we get in a car. Eran is making similar choices with the decisions he makes every day about diet, sleep, exercise, and seeking to minimize stress.

Although my lifetime risk, with a single APOE4 allele, is theoretically smaller than my son's, my age makes me a more likely victim in the coming years. I'm not letting fate decide by itself, though. Even now, in my eighties, I generally exercise one to two hours a day. I do a set of 200 push-ups and 100 sit-ups each morning. I do a lot of stretching and balancing exercises, and I run, cycle, and/or walk briskly several miles most days. I also follow a diet intended to keep me healthy in all aspects of life, with lots of whole grains, legumes and nuts, plenty of green leafy vegetables, a good helping of fruits and berries, and plenty of water every day. I practice intermittent fasting and am beginning to do BrainHQ most days. I enjoy a healthy social life with friends and colleagues, attending sporting events, concerts, and lectures whenever I have the chance. I read lots of books, both fiction and nonfiction. I collect art. I work long hours as a human biologist. Am I immune to Alzheimer's? Of course not. But given my increased risk and the heartbreaking experience of watching my children slowly lose their mother, it seems to me that it would be foolish not to do what I can to protect myself.

How early should someone with an increased risk of Alzheimer's begin such interventions? That's a question our family is asking right now. Because Eran has two copies of the APOE4 gene, his two children, both girls, Sidney and Maia, must have at least one bad copy each. When should they be told? When should they be tested? When should they be encouraged to adopt the modified multimodal type of regimen, with exercise, appropriate diet, low alcohol and no cigarette consumption, and careful monitoring of relevant blood chemistries?

My conviction is "sooner rather than later." When it comes to developing lifetime habits, the earlier one starts, the easier it is to succeed in establishing them.

A Painful Choice and a Beautiful Hope

Several years ago, I made the decision to stop all of the ineffective drugs that Valerie had been on since the beginning of her disease. They clearly weren't helping, but doctors had long advised to stay the course, reasoning that she may go through a painful withdrawal if the drugs were discontinued. For a time, I consented. Now, I wish I hadn't, for it was soon after ending the treatment regimen that Valerie was transformed, at least for a time, from a somnolent and nonresponsive patient with her eyes shut much of the day to someone far more alert and responsive.

This certainly hasn't cured her disease. It was simply a small victory that felt like a very big victory. The long-term prognosis is no different now than it was before. Clearly, though, sometimes no drugs are better than ineffective drugs, and I do wonder what aspects of Valerie's mind or personality might have been retained or restored if we'd ended the drugs sooner. I have also often wondered how things might have been different for Valerie, for me, and for our children, if I had known in 2005 what I know today. If we had known, just a few years ahead of time, that Valerie was at risk, what choices would we have made? If we could have kept watch for the minute chemical signs that can warn us that a wellness-to-disease transition is under way, what options would still have been on the table? If we could have started her on a preventative or responsive multimodal therapy program, what would our lives look like now? Would the woman I have loved for all these years still be at my side? There is no way to know, of course, and perhaps these sorts of ruminations are unproductive. But the idea that an individual multimodal therapy can be a preventive therapy for a high-risk Alzheimer's patient is compelling.

I cannot turn back time, but perhaps I can help spare someone else the suffering that Valerie, our children, and I have been experiencing for more than seventeen years. So it is that I have turned a significant part of my professional attention to developing new insights and approaches to Alzheimer's disease. I believe very strongly that this work will help many people in the future.

If this work were to become part of Valerie's already shining legacy of contributions to science and medicine, I have a feeling she would be very pleased. I only wish I could tell her so and know that she understood.

8

Deciphering Dementia

How a Radical New Understanding of Alzheimer's Disease Is Offering New Hope

The last restaurant dinner we had before sequestering away in response to COVID-19 was with Thomas Paterson, a brilliant computational modeler we've known and respected for many years. Tom brings together two sets of expertise: dynamical systems theory, which he studied as an undergraduate at Massachusetts Institute of Technology, and decision theory, which he studied as a graduate student at Stanford. In the month leading up to our dinner, Nathan and Tom had spoken for hours about the premise of scientific wellness—the idea that, to understand disease, you first have to understand wellness.

The company Tom founded, EmbodyBio, is in the business of creating "digital twins," a simulation based on troves of personal data intended to show how different health decisions and interventions will impact individuals differently. Want to know how a new medication will interact with a drug you're taking, given your specific genetics, phenotype, microbiome, and lifestyle habits? Tom and others working on this exciting frontier of science believe that a digital twin can offer important answers. But digital twins can't yet be built to account for every aspect of human health, so Tom had an important question for us: What problem is most worth taking on?

Have you ever noticed how, often, actors in stage plays, television shows, and movies answer questions in unison? It's an overused trope, because it doesn't happen very often in real life. But on that day, it did.

"Alzheimer's disease," we said at the same time.

A few days later, the restaurant was closed, and Seattle settled in for a long COVID-19 shutdown. Over the next few years, as the world turned its attention to COVID-19, Tom focused on another global scourge. Using the EmbodyBio platform, he worked with us to build out a dynamic model of homeostasis, the processes by which the body (including the brain) maintains its health state. The model was especially focused on the reconstruction of "feedback loops" that are essential for regulating our normal biological processes and making us resilient to outside influences. A commonly understood example would be how our bodies manage to maintain a core temperature of 98.6 degrees even when outside temps are freezing or sweltering. In one instance, you may find yourself shivering; in the other, sweating. But many things happen below the surface to produce those palpable manifestations of the body's adjustment. Such feedback mechanisms to maintain homeostatic consistency exist everywhere in the body and are essential to maintaining our health and keeping us alive. Understanding when and why they break down is thus essential to understanding disease.

After our meeting, Tom and his colleagues Jen Rohrs and Don Breuner worked to connect Tom's world of physiological feedback loops with our world of genes and molecules. Could this integrated model help us fit together the much-studied pieces of the Alzheimer's puzzle, like tau and beta-amyloids, into a coherent whole? Even better, would we be able to use the model to interpret the backlog of Alzheimer's data that were still poorly understood?

It turned out we could.

A New Mechanistic Understanding of Alzheimer's

In the early months of our collaboration, we met regularly with Tom and Jen, along with Cory Funk and Priyanka Baloni, senior scientists in our lab at ISB, to discuss applications of the model in various areas of current research. Later, with Nathan's move to Onegevity/Thorne in 2020 and the subsequent formation of a partnership between Thorne HealthTech and EmbodyBio, neurodegenerative disease researcher Ben Readhead, biochemist Sheena Smith, molecular biochemistry experts Stephen Phipps and Michael Schmidt, and brain health clinical specialist Mary Kay Ross became key contributors. Assembling, testing, and refining the model took unexpected turns.

Counterintuitively, much of the prior research guiding this work wasn't originally focused on Alzheimer's. Of the more than 700 scientific papers used to model development, nearly half were not about, or even cross-referenced to, Alzheimer's disease. They described biological elements that were essential to homeostatic feedback loops in the healthy brain but were also pivotal for understanding how Alzheimer's risk factors could destabilize homeostasis in the aging brain.

One of the benefits of working with such an integrated approach is that the model needs to explain the observations of all these studies. You can't have one explanation to account for the data in one study, and another for another study. Thus, while the individual pieces come from discrete data sets and clinical observations—the systems approach—putting it all together and modeling it quantitatively can illuminate what it means as a whole.

One of the most challenging aspects of studying disease is separating the signal from the noise. When you run an experiment and generate a large amount of data, differences will crop up between the disease samples and the healthy controls. For example, if you measure a "transcriptome," you will get a measurement of the expression levels of thousands of genes simultaneously. The same is true for proteomics and metabolomics. Somewhere buried inside these observed differences are causal factors—but identifying them correctly is a challenge in a complex, dynamic system that asks us to determine which changes reflect negative consequences stemming from the disease and which are appropriate responses of the body in an effort to combat the disease and return the body to equilibrium. This distinction is critically important. Otherwise, addressing a difference between diseased and healthy samples could inadvertently make the problem worse by targeting for elimination something the body is doing to counteract the disease. Systems models can serve as powerful lenses through which to discern these different changes.

So what did our initial focus on modeling how the brain maintains health reveal about the drivers of Alzheimer's disease? What did we learn about the causes, consequences, and compensatory mechanisms?

At its heart, we now believe that the Alzheimer's key disease mechanism tackles metabolic energy deficits in the brain. It appears that homeostatic compensation for neuronal loss, such as that which occurs with negative energy balance, expands the susceptibility of the remaining neurons to these energy deficits. This creates the positive feedback loop at the heart of Alzheimer's, as the loss of the most vulnerable neurons puts added pressure on the next most vulnerable neurons, leading them to die, and onward it goes as the degeneration cascades. Once we accepted this basic premise, we began to see that many of the apparently diverse observations of potential causes of Alzheimer's funnel back (in interesting and surprising ways) to this fundamental issue.

On one level, this makes perfect sense: the brain is highly metabolically active and has massive energy needs. It represents about 2 percent of body mass but consumes a whopping 20 percent of its energy.[1] The brain's neurons and other supporting cells need to maintain their energy—generally quantified through the currency of high-energy adenosine triphosphate, or ATP—to stay alive and enable cognitive functioning. So there is a high demand for energy that must be satisfied by enough energy production. That's a pretty straightforward proposition—mass-energy balance is about as fundamental as it gets in science—but many complex factors play a role in making it happen.

Let's start with what's known as the oxygen perfusion rate, the amount of oxygen available to the cells in the brain. This is not uniformly distributed. While the brain as a whole consumes 20 percent of the oxygen delivered to the body, this oxygen doesn't go everywhere equally.[2] Certain regions of the brain are more challenging to reach; as a consequence, the neurons in those regions are less efficient at producing the energy they need to survive. That's really not a problem during most of our lives, because there is enough excess capacity across the whole of our brains. But our ability to perfuse oxygen—to diffuse oxygen efficiently even to the most remote regions of our body—decreases as we age, and we now know that this loss can contribute to dementia in later life.[3] We also know that the rate of this decline is exacerbated by modifiable lifestyle factors such as obesity and smoking.[4] According to a 2020 report from the Lancet Commissions, which matches editors from the prestigious *Lancet* journal with academic partners to identify and craft recommendations for addressing the most pressing issues in global health, current smokers had a 40 percent greater risk of developing Alzheimer's disease.[5] The reasons for this link are not fully known, but the authors of the report believe the most likely cause is oxidative stress. This makes sense. It is both obvious and well studied that smoking has a profound negative impact on oxygen availability to the lungs and, from there, its diffusion to the rest of the body, including the brain. And if Alzheimer's disease is fundamentally caused by an inability to generate the energy we need to keep our neurons alive and synapses firing, then this would represent a clear, causal, mechanistic pathway and help explain why Alzheimer's disease generally arises later in life, when a natural decline in the ability to deliver oxygen occurs.

One of the biggest factors that slows the decline in—or improves—oxygen perfusion is exercise. Again, this makes sense, as vigorous exercise increases respiration. So it is understandable that exercise has repeatedly been shown to be the single most effective way of improving many of the neuropsychiatric symptoms of Alzheimer's.[6] Most studies have focused on the effects of exercise on people who have already been diagnosed with dementia. But as we now know, clinical diagnoses generally come many years after a disease transition has taken place, by which point the disease has become firmly rooted in multiple systems.

To investigate more proactive interventions, we took our cue from the successful FINGER trial discussed in the previous chapter. This two-year, double-blind, randomized controlled trial included 1,260 people who were older and at risk for Alzheimer's and related dementias but who did not yet have clinical signs of these diseases. Participants in the trial were encouraged to take part in a multimodal therapy program that prioritized exercise, nutrition, cognitive training, and social activities, along with intensive monitoring and the management of metabolic and vascular health. This approach, akin to what we are calling scientific wellness, only without the data-crunching function of AI, was the most successful trial to date, outperforming all previous drug trials in maintaining or even improving cognitive function.[7]

It's important to reiterate that exercise was only one component of the FINGER regimen, but the fact that it was an essential component is tremendously encouraging, because it aligns with our emerging understanding of the role that the effective dissemination of oxygen to the brain plays in Alzheimer's. What's more, while prevailing theories about the accumulation of amyloid plaques and neurofibrillary tangles don't offer an obvious explanation for why exercise helps, the oxygen perfusion hypothesis clarifies why it does. So even though it's not the whole picture when it comes to Alzheimer's, it is a central part of the picture that we can actually do something about, and rather simply.

Another intriguing observation supporting the centrality of oxygen perfusion for preventing Alzheimer's appears through analysis of electronic medical records by a team of researchers at the Cleveland Clinic. Published in *Nature Aging* in 2021, this study showed that the use of Viagra (which increases blood flow in the brain as well as elsewhere) was associated with lower rates of Alzheimer's disease.[8] Whether this effect is causal will need to be tested in a randomized clinical trial; other therapeutic interventions should also be possible under the new paradigm.

Employing the Digital Twins

How do we know who will be most advantaged by the increased flow of oxygen that comes with more intensive exercise or other simple, early interventions that might delay or prevent transitions to Alzheimer's disease and other forms of cognitive decline? A big part of the answer comes from genetics, and a good starting point for that conversation is the gene known as APOE. The APOE4 variant—of which Lee's son, Eran, has two copies—is a major genetic risk factor for Alzheimer's. It has been estimated that individuals with one bad copy of this gene may have a threefold increase in their lifetime risk of developing Alzheimer's. Having two bad copies doesn't just double that risk—it might triple it.[9] The APOE2 variant of this gene, in contrast, reduces the risk of Alzheimer's by 87 percent compared to the most common genotype, which is known as APOE3/3.[10]

To explain why APOE variants can have such a pronounced effect on Alzheimer's, it's helpful to understand one of the gene's primary roles: it codes for a protein that facilitates cholesterol metabolism. And, just as is the case with energy metabolism, a lot of this happens in the brain, which accounts for approximately 20 percent of the cholesterol in our bodies.[11] Interestingly, experimental studies have established different rates in the ability to transport cholesterol exhibited by different variants of the APOE gene. APOE4 carriers are prone to high brain cholesterol acquisition, while research suggests that APOE2 carriers are likely to be somewhat protected from cholesterol accumulation.[12]

When we simulated the effects of these differences in a dynamic model of brain health, the result was a fascinating and revealing picture of how different APOE variants impact the ability of neurons to maintain enough ATP (energy) production to stay alive under decreased oxygenation and increased demand. A key factor for maintaining the feedback loops that coordinate energy support for neurons appears to lie in keeping an optimal level of cholesterol in astrocytes, non-neuronal cells that play many important, neuron-supportive roles for functions in the brain.

To understand how these individual differences might play out over a long human lifetime, we set Tom Paterson's digital twin model into action. The ultimate purpose of a digital twin is to be a stand-in for an actual person. For now, since we don't yet have 360-degree multiomic measurements (i.e., deep phenotyping) for millions of people, Paterson and his team created variations in genetics and physical and biochemical characteristics that represent what could be seen in ten million different patients. Of course, none of these ten million "patient simulations" is a perfect avatar for any single person on our planet, but as one of many indicators that the team got things right, at least from a population health perspective, the simulated patients wound up with the clinical symptoms of Alzheimer's at strikingly similar rates as real-world patients at each age, across the decades of life, for each of the APOE genotypes.

Our findings about oxygen perfusion and the genetic predisposition for cholesterol accumulation are further indications that the prevailing hypothesis about Alzheimer's, which for decades has focused on the buildup of amyloid plaque, is woefully incomplete. If we'd made these discoveries a decade or two ago, we may have been seen as pariahs in this field, but following hundreds of failed clinical trials, the tide has been turning against the conventional amyloid thesis of Alzheimer's for several years.

It's true that efforts to find a "magic pill" for Alzheimer's are ongoing—and most of these still target amyloid, the protein whose aggregation in the brain was one of the first features noted originally by Dr. Alois Alzheimer at the beginning of the twentieth century. One recent and unfortunately harmful example sought to create drugs aimed at a gene called BACE1 (β-site APP cleaving enzyme-1) in the hope of limiting the production of amyloid. These attempted therapies were very successful in lowering amyloid in the brain, but was there any cognitive benefit to patients? No. In fact, multiple trials were discontinued prematurely because of this futility and patient safety concerns.[13] A clinical trial with the BACE1-targeting drug known as Verubecestat showed up to an 80 percent reduction in amyloid in cerebral spinal fluid, along with a modest reduction in plaque measured using PET scans. But the trial was terminated early when it became clear that these reductions weren't coming with any meaningfully beneficial effect on cognition. The drug appeared to be causing a small decrease in patient brain volume, and cognitive symptoms of those with early Alzheimer's actually seemed to be worsening.[14] Yet remarkably, after all these failures, efforts are still under way to develop more BACE1-targeting therapies.[15]

As late as 2021, the Food and Drug Administration (FDA) approved a new monoclonal antibody that targets amyloid called aducanumab, created by Massachusetts-based biotechnology company Biogen and sold under the brand name Aduhelm. This approval was a controversial decision, to say the least. None of the eleven scientists on the advisory panel voted to approve the drug, citing a lack of evidence for efficacy.[16] The FDA took the highly unusual action of overriding the recommendations of the scientists on the board and moving ahead with approval anyway, reasoning that, although it showed no significant effects on cognition, at least it reduced amyloid plaques.

You might be inclined to see this series of events through the lens of craven capitalism—there is money to be made with new drugs intended to stop the buildup of brain plaques, even if those drugs don't work very well for anyone and won't work at all for many people who are suffering from Alzheimer's. And that's a fair suspicion. Biogen's market cap jumped $50 billion in the stock market on the news of the approval, and the price per patient for this drug has been set at $56,000 a year. But why would the FDA care about Biogen's stock price? Even knowing that it wasn't likely to help most Alzheimer's patients, and even recognizing that it was apt to be cost-prohibitive for many sufferers, some patient advocacy groups were clamoring for aducanumab to be approved, desperate for any possible help they might get. It may be that Biogen and FDA officials were simply responding to a frantic cry from a community of people suffering from the disease or caring for those who did.

In whatever light you view this situation, one thing is incontrovertible: the overwhelming majority of Alzheimer's patients won't be helped one bit by this drug. The belief that reducing amyloid must be good for reducing Alzheimer's disease, despite so much evidence to the contrary, has proven persistent because so many researchers have been so sure for so long that amyloid simply must be responsible. But, as it is often said in a line of folksy wisdom dubiously attributed to Mark Twain, "It ain't what you don't know that gets you into trouble. It's what you know for sure that just ain't so."[17] The hunt for new ways of targeting amyloid plaque is another example of looking for your keys under the lamppost because that's where the light is.

Reevaluating the Amyloid Association

Yes, the association between amyloid and Alzheimer's is powerful. But what we now believe is that amyloid is primarily a compensatory mechanism triggered by the brain in an effort to enable the formation of new synapses.[18] Its main function, it seems, is to help maintain cognitive ability once the cascade of neuronal death has been triggered by metabolic exhaustion.

To understand how that works, we need to go back to one of the principles we discussed in Chapter 6—the Hebbian learning theory of activity-dependent synaptic plasticity, which suggests that "cells that fire together wire together." In this model of how neurons wire themselves over time, there is always a tradeoff between too little and too much firing of the synapses. If there is too little background firing, the correlation may not be sufficient to induce new brain wiring. But if there is too much background synapse firing, random noise will swamp out the signal and correct wirings won't be made. So for effective information coding to happen, neuronal firing rates need to stay within a certain "Goldilocks" activity level.[19] And that's where amyloid comes in. As synapses are dying off, the brain secretes amyloid because it induces just enough background "noise" into the system to get nascent synapses "off the bench," creating an opportunity for Hebbian connections and stronger synaptic connections. Thus, amyloid is a tremendously good biomarker, one that indicates there is an underlying issue affecting cognition. But it is a terrible drug target, because you are attacking a response, not the disease. Phosphorylated tau, another molecule of intense study in the Alzheimer's research community, may also be largely connected to the same underlying metabolic mechanisms, in that tau serves as a biomarker of neuron metabolic exhaustion.[20]

Let's put some of these pieces together (Figure 8.1). In this conception, the most vulnerable neurons become metabolically exhausted as we

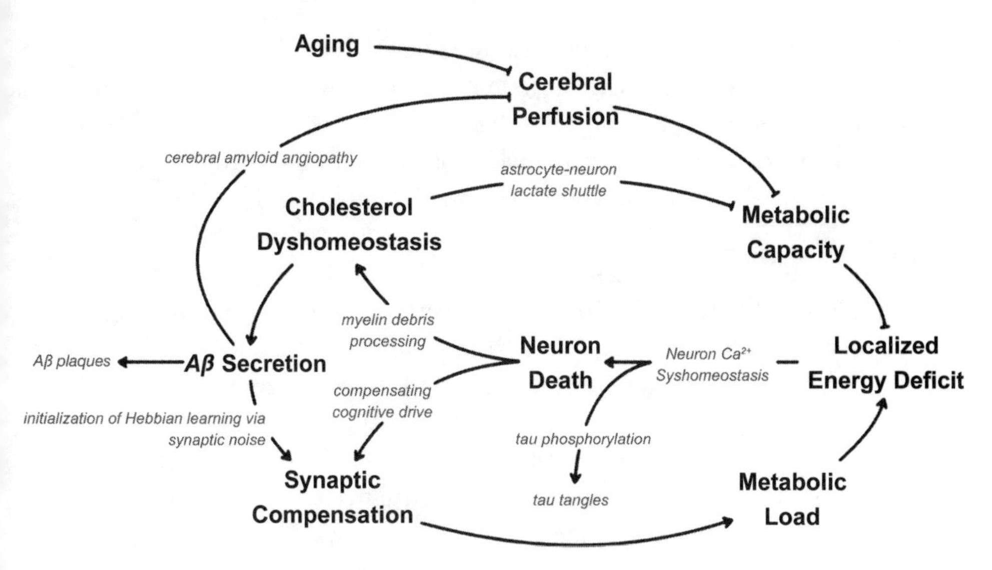

Figure 8.1. Schematic of central mechanisms driving Alzheimer's disease.

age and die due to an inability to satisfy their energy requirements. This results in even greater metabolic demands on the remaining neurons in order to maintain cognitive function. What's more, these remaining neurons have to clean up the debris from the dead cells around them, and especially the cholesterol that was previously sequestered by the now-expired cells. Now, imagine what would happen if every homeowner in your city decided to clear out their attic and basement and set the debris on the side of the road for pickup at the same time. The collection service would be quickly overwhelmed.

That's what's happening in an Alzheimer's-affected brain. This overloaded cholesterol debris processing leads to elevated cholesterol levels, impairing the feedback loops that coordinate energy support and making it harder for the remaining cells to meet their energy needs. This increased energy burden leads to the next most vulnerable cells in metabolic exhaustion dying, resulting in more debris and leading to more cholesterol dyshomeostasis in a vicious cycle that ends with many of the symptoms of Alzheimer's that far too many of us have come to know and loathe.

As is the case in any scientific field, there are always diverse voices and individual scientists who see the problem differently from the mainstream. Ours is not by any means the first critique of the amyloid-centric view that has overwhelmingly dominated Alzheimer's research over the past few decades. Alexei Khoudinov from the Russian Academy of Sciences was an early and important pioneer. He argued in two key papers in 2004 and 2005 that amyloid plaque, far from being the cause of Alzheimer's, was a protective mechanism linked to cholesterol homeostasis.[21] But while it might seem obvious now that the research community should have seen the faults in the amyloid-based fixation and listened to the voices of the naysayers all along, it is important to understand why there was so much consensus in a field full of incredibly smart and dedicated people, because this was more than a simple case of mistaking correlation for causation. After all, it's not that amyloid is only a marker of disease and nothing else. It's more complex than that. As we noted earlier, it does play a causative role in changing the brain's chemistry and cellular networking in ways other than its well-known plaques.

Amyloid also may be fundamental to helping researchers understand a distinct subset of early-onset Alzheimer's disease. Mutations in three genes—amyloid protein precursor (APP), presenilin-1 (PSEN1), and presenilin-2 (PSEN2)—substantially increase the risk for familial early-onset Alzheimer's disease, defined as Alzheimer's before age 65 that in some cases results in dementia arising decades earlier.[22] Each of these mutations has a significant effect on the amyloid compensatory secretion rate associated with the loss of synapses. So, while we do not believe that the formation of amyloid plaques is the central cause of Alzheimer's, we do note that the higher rate of synapse firing induced by amyloid does cost energy, which of course is a central part of our alternative view of the genesis of this disease.

One of the biggest issues that arises downstream of this amyloid-induced increase in synaptic activity is that the increased metabolic load results in a hyperactive hippocampus. Essentially, it "burns hot," and this complex brain structure, which plays a major role in learning and memory, burns out. This is where the first deficits in brain functioning arise in the transition to Alzheimer's. In fact, the rate of energy consumption triggered by this process ramps up significantly—so much so that, when we included this higher energy consumption rate into our models, we saw something that took our breath away: when given these genetic mutations, the age at which our ten million digital twins succumbed to Alzheimer's was shifted decades earlier, to their thirties and

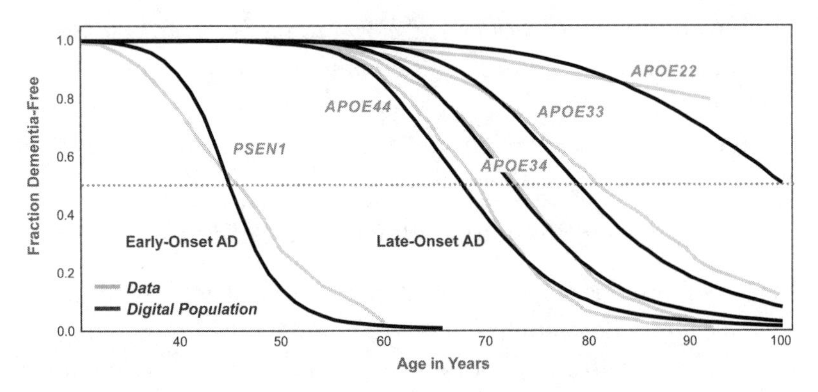

Figure 8.2. Model predictions can capture the different ages at which people will come down with Alzheimer's disease as a function of their genotypes with a high degree of accuracy.

forties, just as is seen clinically (Figure 8.2). The same central mechanism that is primary in the normal progression of Alzheimer's could also explain the early-onset cases of this disease!

Other prominent Alzheimer's genetics can also be explained based on how they affect the central metabolic mechanism. For example, one of the most interesting genes to be studied by the Alzheimer's community in recent years is TREM2. This gene and a number of identified variants are best known to be related to microglial phagocytosis—the process by which bacteria are detected and destroyed by specialized cells in the central nervous system. When combined with two other Alzheimer's-associated genes, APOE and ABCA1, however, TREM2 helps give microglia the ability to sense and offload excess cholesterol—which, as we earlier noted, is a key challenge that comes in the wake of the death of neural cells. Impairment of TREM2 activity conversely results in dysfunctional microglia that are unable to efficiently process myelin debris, made up of the broken pieces of the insulating sheath that forms around dead or dying nerves.[23]

In a remarkable display of functional flexibility, the brain is able to adapt astrocytes—the support-giving non-neuronal brain cells—to aid in the clearance function that is the natural domain of the microglia when the microglia are overwhelmed. But when astrocytes attempt to compensate for microglia impaired by built-up debris, the excess cholesterol impairs metabolic support, further accelerating disease progression.[24] Thus, once again, seemingly diverse effects from genes involved in various biological processes are shown to be interconnected across the same fundamental processes the brain uses to maintain its health. This is the essence of an integrated systems and scientific wellness approach to understanding disease.

Using Blood to Understand Our Brains

There are hundreds of additional gene variants that contribute to one's polygenic risk of developing Alzheimer's, and it will require significant work to contextualize each of these genes under the emerging model. It will also take a lot of data from different brain states. Fortunately, Laura Heath is on the case. You couldn't find a more qualified person for such an endeavor. Heath, who was a senior scientist in our Hood-Price Lab for Systems Biomedicine at ISB, is a public health geneticist with expertise in epidemiology, genomics, bioinformatics, bioethics, and the analysis of complex data sets, and what she has been able to do with the Arivale data is inspiring. She is looking at the manifestation of genetic risk in blood analytes through the lens of all the known Alzheimer's-associated gene variants in the Arivale population for clues about the biology of high-risk individuals with no presenting symptoms. Such efforts often begin with acts of calibration. For starters, Heath confirmed the finding that those with APOE4 tend to have elevated LDL cholesterol levels in their blood. She was also able to show how this relationship changed dynamically across the decades of life. But cholesterol is just one potential marker associated with Alzheimer's, and Heath also found about two dozen other proteins and metabolites that were altered by Alzheimer's-associated genes. She noticed that the effects in the blood of many of these genetic variants were different between men and women. This could offer clues as to why women contract Alzheimer's at a much higher rate than men.

Heath's work enabled us to look for and identify more of these early indicators in patients with different genomic and phenomic makeups—pointing to a diversity of potential paths for early intervention. It is noteworthy, however, that the strongest of all the signals is still the association between APOE4 and elevated blood cholesterol. This once again points to the central mechanism we have described in this chapter.

Additional evidence for this central model of Alzheimer's disease comes from the effects of drugs that could touch on the key mechanism of cholesterol trafficking: the class of lipid-lowering medications known as statins, and in particular lipophilic statins, which have an affinity for fat and can more easily pass through the blood-brain barrier, the semi-permeable layer of cells that keeps many of the chemicals found in circulating blood away from our neurons.

To investigate the association between brain health and these types of statins, which are commonly known by brand names such as Zocor, Lipitor, and Altoprev, a research team led by Daniel Silverman at University of California Los Angeles (UCLA) followed the health trajectories of 300 older adults with mild cognitive impairment—the first clinical signs of dementia—over eight years. Those who were using lipophilic statins were more than twice as likely to develop dementia over eight years.[25]

What really grabbed our attention in this study were the PET scans, which helped illuminate the areas of the brain that died earliest in the progression of Alzheimer's disease. These were the very same regions that we and other researchers have identified as being most affected by "hypometabolism," otherwise known as a slow or sluggish metabolism, and thus as commonly impaired through disrupting cholesterol homeostasis in individuals with the APOE4 variant.[26] There was one way, however, that this study at first seemed to be at odds with our emerging hypothesis about the progression of Alzheimer's disease: in targeting cholesterol, lipophilic statins might actually be expected to play a beneficial role in reducing dementia. Indeed, there were previous studies suggesting lipophilic statins might be associated with a *reduced* incidence of Alzheimer's.[27] So what was going on?

Here is where the difference between doing your research on the internet and deep science comes into focus. Headlines aren't always helpful. Even well-written research abstracts, which are a necessarily abridged description of the scope and results of a study, can lead a reader to conclusions and associations that just aren't true. This isn't because of any willful attempt to deceive or even because the reader isn't "smart enough" to understand. Two scientists with closely shared expertise can read the same abstract and come to very different conclusions about what it means. What often matters most when we try to apply new research findings to our understanding of the world is not what a study purports to have shown but how it was conducted. What were the criteria for being enrolled? What were the data? How were the numbers crunched?

In this case, a really important factor was that those participants being recruited into this study had already entered mild cognitive impairment, which is defined as partial loss of memory or other cognitive traits in people who nevertheless maintain their overall cognitive ability sufficiently to function independently in daily living.[28] So the downward cognitive cascade had already been initiated.

When we incorporated data from the seemingly conflicting papers into our model and then simulated the effects of lipophilic statins on their known target for ten million digital twins, a critically important subtlety emerged. In digital patients who had already entered into some degree of cognitive impairment—those who had left the state of transition and were now in a state of disease, even if only a "mild" one—our results were consistent with those of the UCLA team. The lipophilic statins were associated with an increased progression into worse states of dementia. But lipophilic statins taken in advance of mild cognitive impairment, usually when someone is in their fifties or sixties, were associated in our digital patients with a delay in the onset of mild cognitive impairment followed by a sharper drop-off between mild impairment and full-blown Alzheimer's later in life. In essence, the administration of lipophilic statins changed the shape of the "cognitive curve" throughout life, providing more life at fuller cognitive capacity but a steeper and shorter ultimate dive into dementia once mild cognitive impairment was reached.

This isn't a trade-off that everyone would view in the same light—the long-term result remains frightening—but it is a choice we think many people would very much like to have. It may also have tremendous implications for healthcare costs, as caring for Alzheimer's patients over many years, and sometimes many decades, exacts a huge toll on individuals, families, and society, with estimated annual costs of $500 billion a year in the United States alone.

A New Direction for Alzheimer's Research

There is a clear and promising path for attacking Alzheimer's disease in the coming years—a path that can restore optimism to an area of research and medical care that has been increasingly lacking in hope. This path uses the powerful tools of systems biology to analyze and understand the biological networks that are impacted during the transition and progression stages of Alzheimer's disease. To this end, we have initiated several clinical trials using individuals at early stages of the Alzheimer's disease spectrum. Spearheaded at ISB by Jared Roach, a senior research scientist in our lab, these are prospective randomized controlled trials, or RCTs, the gold standard for proof in translational medicine studies. For these trials, we selected elderly individuals who have just begun to show evidence of transition to Alzheimer's disease or other forms of dementia and performed deep phenotyping to obtain blood biomarkers that explain key wellness and disease transitions. These biomarkers can then be used in wide-scale population screening to identify individuals at the onset of disease before their symptoms have become observable. Over time, analysis of individual data clouds will lead to the identification and definition of different contributors to, and subtypes of, Alzheimer's disease. Most important, these data will allow us to identify data-informed multimodal intervention strategies for personalized care and disease reversal. We believe such a strategy will be broadly applicable, and not just for some who happen to respond to one particular treatment. Through personalization of multimodal therapy possibilities, we aim to help everyone who is at risk of this disease, each in their own way.

Does that sound like an awfully bold vision? Well, it is. And we need convincing clinical trials to support this multimodal approach. We were fortunate in this regard that the Seattle-based Providence St. Joseph Health system had an executive, Mike Butler, who was sympathetic to our cause, having experienced the tragedy of Alzheimer's disease in his own family. Thanks to his leadership, Providence was willing to provide $8 million to initiate multimodal therapeutic clinical trials. As of this writing, we have two multimodal Alzheimer's clinical trials and one observational study under way.

The first of our two trials, Coaching for Cognition in Alzheimer's (COCOA), is a collaboration with William Shankle and Hoag Hospital in Orange County, California. COCOA aims to evaluate the effectiveness of adding multimodal behavioral lifestyle coaching to conventional Alzheimer's disease treatments. These coached interventions are informed by multiomic deep phenotyping data similar to those obtained with Arivale. Interventions include diet, exercise, and cognitive training provided by BrainHQ.

The second trial, Precision Recommendations for Environmental Variables, Exercise, Nutrition Training Intervention to Optimize

Neurocognition (PREVENTION), is being conducted in conjunction with the Pacific Brain Health Center, a Santa Monica–based affiliate of Providence St. Joseph Health.[29] PREVENTION is led by a team of physicians and scientists including Drs. David Merrill, Jared Roach, and Jennifer Bramen. Like COCOA, PREVENTION focuses on people suffering from mild cognitive impairment and seeks to evaluate the efficacy of a data-driven multimodal approach to therapy, including lifestyle coaching. PREVENTION aims to validate and confirm the findings of COCOA, as well as to specifically study multimodal interventions in individuals already confirmed to have pathophysiology such as amyloid in their brain. It includes health coaching, dietary counseling, group cognitive and physical exercise classes with a certified personal trainer, and computer-based neurocognitive training. One objective is to assess the impact of this multimodal intervention on the patients' cognition. Another objective is to generate dense, dynamic data clouds across the patients to look for clues to what differentiates those who benefit substantially from the treatment from those who do not.

Finally, we are working with a network of physicians coordinated by Dr. Mary Kay Ross, a leading brain health medical practitioner and now chief medical officer of Thorne HealthTech, on an observational study to track clinical outcomes for all of her consenting patients, as well as patients across a number of leading brain health practices so that we can do a proper assessment of the trajectory of everyone involved in these multimodal therapies and assess the predictive ability of the digital twin simulations of our Alzheimer's computational model for assessing the effectiveness of therapies on an individual patient basis. By analyzing de-identified data for all patients—and not looking only at the cases with good outcomes—we will be able to evaluate the efficacy of the program against typical norms. In turn, the deep phenotyping data created by this effort will provide feedback and enable us to refine the valuable personalization of these approaches, giving Dr. Ross even more tools to help her patients.

Through Thorne HealthTech, we are also doing a pilot program with six physicians to get feedback on the design of a brain health program informed by the dynamic model of brain homeostasis and an AI system that encapsulates the current state of knowledge around wellness-based disease prevention. Our goal is ultimately to create a broad network of physicians who are applying these approaches in their practices, with de-identified data sharing that will allow us to track progress, recognize which parts of the program are most effective, and focus preventive approaches to be as effective and economical as possible. These collaborations will also give us insights into how physicians can treat individual Alzheimer's patients more effectively.

One final thing we are working on is following with genome/phenome analyses the paths of individuals at high risk for Alzheimer's disease in their fifties and sixties who don't show any clinical signs of the disease. The idea is to pinpoint blood biomarkers that could detect the wellness-to-disease transitions up to ten years or more before the disease would normally be recognized so that "preventive" multimodal strategies may be initiated to prevent clinical disease. These initiatives are at the vanguard of how much chronic disease can be approached through scientific wellness and scalable efforts that are predictive, preventive, personalized, and participatory.

This will all take time. But that doesn't mean that we should wait to take action against Alzheimer's. We do not need clinical validation for patients and their families to take the kinds of steps that are being recommended by multimodal brain health physicians, the FINGER trial, and simple common sense. Many of these recommendations are straightforward, safe, and literature-backed, and they often involve activities that coincide with wide-ranging outcomes that are generally healthful. And let's face it: there is little to lose. Drugs have failed to work for Alzheimer's patients and expose them to significant costs, and even the most commonly prescribed ones have little upside and a wide range of noxious side effects.

Alzheimer's as a Model for Other Diseases

Can we dramatically reduce the scourge of Alzheimer's disease in our lifetimes? We believe so. And the sooner we arrive at that momentous occasion in medical history, the better. A victory over dementia would alleviate immense anguish and suffering. But that's not all it would do. We think our work on Alzheimer's could be a model for thinking about other chronic diseases. Signs are already mounting that other neurodegenerative diseases—such as ALS, frontotemporal dementia, and at least one form of Parkinson's disease—may also respond to adaptive multimodal therapy. This is an exciting area to be explored, and the reason we told Tom Paterson that Alzheimer's would be a tremendously positive target for his initial work with us.

At some level, this all makes sense. Many diseases are triggered by lifestyle choices, hormonal fluctuations, and environmental toxins. Many have different paths, known as subtypes, by which the end stages of the disease are reached. Each will certainly require its own unique multimodal strategies for therapy. And when we identify interventions that work in one, it is likely we will have come upon interventions that also work in others. For any of this to be effective, we have to fundamentally change the way we do medicine. We have to focus on wellness, vigilantly screen for wellness-to-disease transitions, and identify treatments that can reverse disease at the earliest transition point, long before it is traditionally symptomatic.

We have a long way to go. But we've traveled a great distance on a path that has gotten us almost nowhere. The prospect of embarking on a new journey should renew our sense of purpose. We hope in the not-too-distant future there will be no more Valeries suffering through seventeen years of progressive mental decline.

9

Cancers at a Turning Point

How New Paradigms Point toward Eventual Cures

In 1971, President Richard Nixon launched a "war on cancer," with the goal of ending cancer by the bicentennial. He was proposing that cancer could be cured in five years. It sounds ridiculous now, but fresh off the success of eliminating polio in the United States and landing a man on the moon, such an ambitious goal must have seemed achievable. Fifty years later, cancer remains one of the top killers in America and much of the world. As of 2019, the risk of developing a cancer in the United States by the age of 74 was 20 percent.[1] Just about every one of us has lost a friend or family member to this scourge.

It's not that we haven't had some successes. Pediatric cancer survival rates are significantly higher than they were in the 1970s, due in part to the greater effectiveness of chemotherapy in the young, and lung cancer rates have plummeted in recent decades, though this is mostly because, after multiple studies exposed the link between smoking and disease, we did a better job of preventing it. We have lowered the death rate from lung cancer more than from any other form of the disease, a testament to the power of identifying causal triggers and implementing preventive measures.

When it comes to overall cancer incidence and mortality, though—especially if you take lung cancer out of the mix—the news hasn't been great. The single-drug, single-target, disease-fighting paradigm that dominated healthcare in the past century has not offered dramatic returns for cancer. So while overall improvements in life spans are occurring, they are happening incrementally. From 2001 to 2017, there was an annual decline in cancer deaths of 1.8 percent per year in men and 1.4 percent per year in women, and "incidence rates are leveling off among males and are increasing slightly among females."[2] We can and really must do better.

The obvious place to begin a discussion of the barbarity of cancer treatments is with chemotherapy, which exploits one of the most fundamental things we know about cancers—that they grow quickly. Chemo treatments were designed to kill any and all fast-growing cells, launching a deadly race in an effort to save patients. The collateral damage can be extensive. When we kill all of our fast-growing cells, our hair falls out and the lining of our gut is destroyed, resulting in nausea so severe that some patients describe it as worse than the cancer itself. Making matters worse, chemo is not all that good at killing cancer cells, so recurrence has been an unrelenting problem.

Azra Raza, a physician and researcher at Columbia University who has spent her life fighting acute myeloid leukemia, painted a bleak and moving picture of the "ghastly toxicities" of cancer therapies in her book *The First Cell—And the Human Costs of Pursuing Cancer to the Last*: "With minor variations, a protocol of surgery, chemotherapy, and radiation—the slash-poison-burn approach to treating cancer—remains unchanged. How good are the solutions we offer if we constantly have to ask ourselves whether the cancer or the treatment we prescribe will kill the patient?" And yet, she notes, "'Stay positive' is the refrain, as if it were a sin to voice the intense pain and suffering of cancer patients."

We do not want to denigrate the work that has been done or the lives that have been saved over the past half-century of this so-called "war on cancer." But can we really call this fight a success? We don't think so—and we are not the only ones.[3] So when President Barack Obama and Vice President Joe Biden called for a "cancer moonshot" in 2016, many people in the medical community simply shrugged. The same was true when President Biden renewed this call in 2021.

We believe it's time to stop shrugging.

The Vastly Diverse World of Cancers

We tend to speak of cancer in the singular, but that is not quite right. The proper way to think of it is in the plural, for cancers are highly heterogeneous, varying from person to person, tissue to tissue, and even cell to cell within a single tumor. While there are certainly shared features, each cancer is different, driven by a unique set of mutations and environmental triggers. This is one of the key reasons we see highly divergent responses to the so-called targeted therapies.

Words matter. They frame our thinking and lock us into conceptual parameters. Traditionally, the word "cancer" is preceded by a word describing a tissue of origin. We have "lung cancer" and "breast cancer," "skin cancer" and "colon cancer." But this is also a faulty way of thinking. For *where* something begins is often far less important than *how* it begins, and cancers begin as the result of genetic mutations that occur in cells over the natural course of life.

A mutation may occur in the EGFR gene, which provides instructions for making a receptor protein that forms a connection point between inside and outside a cellular membrane. Or in the BRAF gene, which codes for a protein that helps transmit chemical signals to a cell's nucleus. KRAS mutations impact a gene that plays a role in a vital cellular signaling pathway. All of these types of mutations can occur in breast tissue, lung tissue, colon tissue, and so on, and the combination of mutations in a particular patient appears to be a far more meaningful starting point for identifying appropriate interventions than the originating tissue type of the tumor.

Until recently, it was relatively easy to see where a cancer was and much harder to see how it had begun. Today, advances in genetic sequencing of both the tumor and the patient's baseline genome have made it possible to identify these specific mutations with much greater accuracy. Single-cell analyses can even show us how different types of mutations are dispersed within a growth that we would traditionally think of as a "single tumor." Tumors have enormous cellular heterogeneity. This is a vital point when considering therapy.

Knowing the location of a tumor is important, but treating a cancer based solely on its location is like fixing up a house based on its zip code. It's perfectly possible, in a neighborhood of older homes where galvanized steel pipes are a common feature, that the plumbing of any one randomly selected home may need to be overhauled. But until you look inside any given home, you won't know for sure if the most pressing need for repair is the plumbing, electricity, roof, something else, or—most likely—some combination of issues. Just as we would assess each house on its own terms, we should investigate each cancer based on its individual mutational patterns, regardless of where in the body it happens to be. This is the start of a precision medicine approach to cancer.

There are so many mutations in most cancers that it can be very hard to identify which ones actually "drive" a tumor's growth. For some of those driven by mutant proteins, drugs exist that can block their activities and "precisely" kill the tumor. This is termed precision cancer medicine. Interestingly, tumors from different tissues may sometimes share the same driver mutation and hence be treated effectively by the same drug. In some cases, there can be spectacular, albeit generally temporary, results, leading to an apparent disappearance of the tumor. Precision cancer medicine has two major limitations, however. First, when the tumor returns, often within a year or less, it is frequently more resistant to therapy than before.[4] Second, these drugs are antibody-based and terribly expensive, generally costing more than $100,000 per year. While they tend to be more targeted and thus have fewer side effects than chemotherapy, they are not without risks, including high blood pressure, bleeding, poor wound healing, blood clots, and kidney damage. This leaves a patient with a difficult choice: spend your life savings to live perhaps one extra year, often with a seriously compromised lifestyle, or let nature take its course.

The longer an initial cancer is allowed to exist, the greater the chance it will become a combination of mutation-driven, disease-perturbed networks, and the harder it will be to cure. Researchers in our lab at ISB worked with Dr. Terry Van Dyke at the National Cancer Institute to study a transplanted brain tumor in mice from its initiation to its final end stage, demonstrating that the number of disease-perturbed networks went from zero to eighteen over eight weeks.[5] By the time most clinical diagnoses are made, tumors are complex, widespread, and resistant to many therapies—because a therapy that might work to solve the problem in one disease-perturbed network isn't necessarily going to work on another.

Only by applying a systems approach to medical research and care can we hope to decipher the disease pathophysiology and stratify cancers into appropriate molecular subtypes to develop personalized cancer therapies. Thankfully, that work has started, and the medical community embraces it a little more every day. Personalized or precision medicine centers now exist at nearly every major academic medical center in the United States, including at Duke, Johns Hopkins, Columbia,

Weil Cornell, the University of Pennsylvania, Harvard, Stanford, and many more. Not only that, they can also be found within most of the top medical centers in Europe, Russia, Japan, Korea, and China. One important recent step in the right direction is the use of combination therapies informed by the unique factors that drive each individual's disease, not unlike the popular three-drug cocktail that revolutionized the treatment of HIV, which works better than any one of its component drugs on their own. (It's worth noting that there was great resistance to the idea of abandoning the one-drug paradigm favored by pharmaceutical companies, insurers, and many doctors. It took a lot of community activism from patients themselves for triple-drug therapy to become the HIV treatment standard.)[6]

Some cancers may be overcome with combination therapies, incorporating a blend of chemical, radiation, and immune agents to optimize success—guided by an increasingly well-rounded understanding of the cancer drivers for each patient. The challenge is multifold, but two factors are paramount: choosing the right combination and dosages of therapies, and administering them at the right time. Today, we are starting to have success on both fronts.

Take, for instance, the use of all-trans-retinoic acid and arsenic trioxide, which have proven to be highly effective for the treatment of acute promyelocytic leukemia, or APL.[7] This powerful combinatorial therapy has transformed one of the deadliest leukemias into a treatable disease with a 90 percent survival rate. The coming years will see APL become even more treatable as personal data clouds offer a clearer view of who is at risk and as advanced screening tools provide easier, faster, and cheaper ways to search for and identify multiple markers for this disease and its predisease transitions. While just a few years ago even the most forward-thinking cancer doctors in the world had no way of helping those at risk, or even knowing who was at risk until symptoms appeared, physicians can now have a far better idea of each patient's individual risk profile. Some have used that knowledge to hone their search for the signals that come in transition rather than waiting for signs of full-fledged disease. One exciting approach uses radar to detect infinitesimally small changes in the size and shape of blood cells that may preclude widespread disease; the approach was reported by researchers in Hungary, an example of how far these innovative changes extend beyond the United States.[8]

This, in our view, is a true twenty-first-century approach to cancer—treatment that starts with prediction and prevention. But before getting there, let's talk about the present, since there is another area of vast therapeutic innovation that has become the source of newfound hope—and it's happening right now, before our eyes.

The Promise of Immunotherapy

Cancer immunotherapy takes advantage of an aspect of our physiology that we have long known about and tried to tap into, but have not been able to harness to its full potential: most cancerous cells are routinely cleared out by our immune system, whose fundamental role is to distinguish our own cells from antigens, the toxins or other foreign substances found on viruses, bacteria, and tumors that induce an immune response in the body.

T cells and B cells, among the basic building blocks of our immune system, play an important role in defending our body against these invaders. There are billions of these white blood cells in the human body. Each T or B cell expresses a single molecular species of receptor that is responsible for recognizing one foreign molecular pattern. It has been programmed to keep watch for a single kind of invader. Collectively these T and B cells express millions of different receptors, allowing them to detect almost the entire universe of foreign molecular patterns or antigens. When these cells recognize a foreign pattern, they trigger cellular division and clonal expansion of the specific immune cell type that can kill the foreign invader. This process works exceptionally well to clear malignant mutations—so much so that, in most cases, we never see or feel those mutations. They mutate fast enough to get the attention of our cellular sentries, a tiny skirmish transpires, and the cancer is destroyed.

Once in a while, however, this doesn't work out. A mutated cell manages to evade detection or overcome the sentries. It divides and divides again, and what starts small can get out of hand quickly. Soon, we have full-blown, life-threatening tumors.

Immunotherapies can be thought of as "training" the sentry T and B cells, helping them to become even better at identifying and fighting invading cells. If that sounds a bit like the premise behind vaccinations, it's for good reason. The principles are similar, only as immune-oncology pioneer Lloyd Old observed as early as 1977, "there is something unique about a cancer cell that distinguishes it from a normal cell."[9] At the time, Old didn't know what that unique quality might be, but over the next half-century, researchers began to identify some of the treacherous tricks cancer cells use to survive immune attacks. Sometimes, they undergo genetic changes that mask their appearance, making them invisible to the immune system. Other times, they develop a shield of proteins that shuts down immune cells or lets them hijack and hide behind surrounding healthy cells.

It took a long time for researchers to understand these tactics, and years more to find ways to counter them. Then came the equally challenging task of overturning the entrenched view that "slash-poison-burn" was the best way to tackle cancer. But in the past decade, there has been an explosion of immunotherapy research, prompting tremendous changes in clinical practice. Old's prediction that there would one day be a fourth kind of cancer therapy—immunotherapy, along with surgery, chemotherapy, and radiotherapy—has finally come true.[10]

Immune Checkpoint Therapies

Lee spent the first forty-five years of his career as a molecular immunologist, probing the gene structures of immune-cell receptors. He looked deeply into the genetic mechanisms that generated millions of different B cell receptors in the immune response and was among the first to isolate T cell receptor genes and study their genetic mechanisms for diversification. He also was among the first to isolate genes that encoded the receptors of the major histocompatibility complex that present fragments of antigens to T cell receptors so they can recognize foreign patterns. And he developed a series of instruments that allowed the immune response to be studied more effectively.[11] In these capacities, he interacted frequently with many of the pioneers of immunotherapy: James Allison, Tasuku Honjo, Irv Weissman, and Steven Rosenberg. A major turning point came in the 1990s, when Allison, then a professor at the University of California Berkeley, observed that a T cell protein called CTLA4 functioned as an immunological "brake," preventing immune cells from reacting to antigens (toxins and malignant foreign cells). Allison was excited by the idea that, if he could somehow disengage the brake, it might unleash T cells to destroy cancer cells. Sure enough, in

1996, he demonstrated that an anti-CTLA4 treatment that blocked the functioning of this "brake" would cure mice of a cancer that would otherwise have been fatal.[12]

Around the same time, a group led by Tasuku Honjo, a professor at Kyoto University, discovered another T cell surface molecule that was also an immunological brake, which came to be called programmed cell death protein 1, or PD-1. As the name suggests, this molecule mediates cellular death, known as apoptosis, in different types of T cells—a process that helps hone immune responses. Honjo went on to develop an antibody that could block PD-1 activity and show that it could also cure an otherwise fatal cancer in mice.[13]

These two advances—later recognized with a 2018 Nobel Prize—led to a new type of immunotherapy now called immune checkpoint therapy, in which drugs are used to block proteins that keep one's immune responses from effectively identifying and attacking cancer cells. This strategy does not work with every cancer, but the exciting thing about immune checkpoint therapy is that, when it works, it really works—and it can have transformative effects for patients with hard-to-treat cancers such as melanoma, multiple myeloma, and renal cell carcinoma.

Another form of immune checkpoint therapy showing great promise is anti-CD47 therapy. The molecule CD47 was first discovered on tumor cells by Irv Weissman at Stanford University. (Lee met Weissman as a senior in a Montana high school, and they have been close friends ever since, collaborating on experiments and sharing a Montana vacation ranch.)

Anti-CD47 therapy targets the CD47 molecule, which has the ability to block macrophages, members of the body's natural immune system, from ingesting cancer cells. CD47 is something of a "don't eat me" signal. In 2013, Weissman and his team developed an antibody that blocked its ability to function and showed that this antibody enabled macrophages to consume several types of mouse blood tumor cells.[14] Their research later demonstrated that many types of human tumor cells displayed similar signals that could be blocked by anti-CD47 therapy. In principle, this approach may be effective for a broad spectrum of cancers, most of which have CD47 molecules on their cell surfaces. This opens up the intriguing possibility that, when fed the right information, the immune system can be trained to target and kill any of these tumor cells that have been instructed to take down their "don't eat me" signs. Weissman has validated this approach for a variety of different cancers and created the company called Forty Seven, purchased by the pharmaceutical company Gilead, which offers a major new immunological approach for many different types of cancers.

Immune checkpoint therapy isn't the only immunotherapy that promises to transform our ability to treat cancer. Another promising tactic is adoptive cell therapy, in which a patient's own immune cells are used as treatment for cancer—either by "accelerating" their natural defenses or editing this defense system to make the cells more effective at recognizing specific mutant tumor antigens. This approach goes back to the 1980s, when Steven Rosenberg, a close friend of Lee from his student days at Johns Hopkins Medical School, was working at the Cancer Surgery Branch of the National Institutes of Health and pioneered the use of the interleukin-2 (IL2) hormone to amplify T cells purified from tumor cells. First, he showed that tumor-infiltrating T cells could be extracted from mouse tumors, amplified, then given back to the animal to trigger the destruction of the tumors. Then, in 1988, Rosenberg reported that his lab had taken T lymphocytes that were derived from the body of a cancer patient, amplified them, and returned them to the patient, where these "leveled-up" white blood cells went to work destroying metastatic melanoma tumors.[15] His lab then pioneered the use of this adoptive T cell therapy in humans. Every patient was given their own amplified immune cells to avoid anti-self-immune destruction. These therapies appear to work best in tumors that are highly mutagenic (with lots of tumor mutations), which is generally true of those cancers caused by environmental carcinogens, such as melanomas from sun exposure or lung cancers caused by smoking.

In recent years, Rosenberg has balanced roles as chief of surgery at the National Cancer Institute in Bethesda, Maryland, and professor of surgery at the Uniformed Services University of Health Sciences and the George Washington University School of Medicine and Health Sciences. He and his collaborators at these institutions have further demonstrated the unique nature of individual killer T cell receptors, the cell-surface proteins that react only to their own mutant antigens. Recently, they have engineered T cells to have receptors that uniquely react with their own mutant tumor proteins to kill the corresponding tumors.[16]

The president of ISB, Jim Heath, previously an endowed professor at the California Institute of Technology, is an expert in immunotherapy and chemistry. Fascinated with the problem of profiling the activity of T cells, Jim turned to microfluidics to study forty or so immune-related proteins in thousands of T cells from a single individual to better understand their unique immune system activity. A spinoff company, IsoPlexis, employed the technology he developed and now routinely assesses immunotherapy patients. Jim is also working with Nobel Prize–winning immunologist David Baltimore and fellow physician-researcher Antoni Ribas to develop personalized cancer therapies, called NeoTCR-T cell therapies, through the company PACT Pharma. This approach designs T cell receptors specifically for each patient's individual cancer antigens, places them into the patient's own T cells in vitro, and then delivers the engineered T cells back to the patient. Their specific antigen patterns are used to identify and eradicate the cancer from within.

These are therapies that have been uniquely designed for each patient—the essence of personalized medicine and, we believe, a blueprint for what is to come.

The next steps include whole-genome sequencing and RNA sequencing of tumor and normal tissue to help identify all mutations expressed by the cancer in a given individual. Self-generated white blood cells that are reactive to these mutations can be isolated and amplified for treatment. Alternatively, gene editing techniques such as CRISPR-Cas9, for which Emmanuelle Charpentier and Jennifer Doudna won the 2020 Nobel Prize, can be used to engineer the appropriate T cell receptor genes, thus generating super T cell killers. The future of adoptive immunotherapy is very exciting.

Vaccinating Cancer

Finally, there are treatment vaccines, which, in the context of immunotherapies, are used not to prevent cancer but to fight it once it has taken hold. These vaccines are often patient-specific and can be made from a person's own tumor cells. Alternatively, they may be created with general antigens that are cancer-type specific, but in that case one has to worry about autoimmune reactions arising from the fact that general tumor antigens can also be found on normal cells. Cancer vaccines to date have had a very checkered history.

All of these immune system retraining approaches are fraught with peril, for the immune system must maintain a constant state of balance. Too little engagement can result in a patient's body being overrun by infections and tumors; too much can lead to autoimmune diseases and other chronic problems. Over-tipping the scale can result in chaos. Still, in a few short years, cancer immunotherapy research has produced results that far exceed those achieved through other tactics over the course of decades.

A report by researchers at Imperial College London published in *Frontiers in Oncology* in 2018 provides a fascinating glimpse of recent improvements in immunotherapies.[17] The team performed a composite analysis across different interventions, trials, and cancers to produce a schematic comparison of outcomes. It estimated that, while conventional therapies for some of the hardest-to-treat cancers resulted in five-year survival rates of less than 10 percent, the first wave of modern immunotherapy (exemplified by anti-CTLA-4 ipilimumab and the therapeutic vaccine Sipuleucel-T) raised the survival rate to nearly 20 percent, and second-generation immunotherapies (like anti-PD-1 nivolumab and pembrolizumab, and anti-PD-L1 agents such as durvalumab and atezolizumab) doubled that efficacy to nearly 40 percent. For some cancers, combination immunotherapy has raised the five-year survival rate to an amazing 70 percent.[18] When this practice was bolstered with tumor sequencing, precision tracking, and a careful control of dosing, timing, and duration, the survival rate was even better.

Given the slow pace of improvement in cancer therapies over the preceding decades, such leaps forward since the first immunotherapy trial in 2015 are simply astonishing. And we have likely just dipped a toe in the pool of what immunotherapy can offer. The promise of immunotherapies has rejuvenated cancer research. It has also driven the development of new cancer technologies, strategies, and companies. Between 2017 and 2021, the number of immuno-oncology drugs in development grew by a staggering 233 percent to over 4,700.[19] A report from the Cancer Research Institute identified a 60 percent increase in the number of academic and research groups actively conducting immunotherapies between 2017 and 2019, leading to the approval of thirty-one new drugs by the FDA. The authors wrote, “One would expect that more paradigm-shifting therapies are on the way.”[20]

They most certainly are. Underlying this growth is a newfound confidence in the effectiveness of the more personalized approach to healthcare we’ve been calling for throughout this book. Immunotherapy is one of the most promising and widely embraced examples of a systems approach to disease. It highlights the benefits of what is sometimes called “N = 1 medicine” (treating each patient individually)—a powerful new direction in cancer treatment and, ultimately, the treatment of all diseases.

Amazing as it may seem, there are three additional new approaches to cancer detection and therapy: the discovery of blood diagnostic biomarkers, the deployment of new data-driven clinical trials, and a growing ability to detect cancer at its earliest transition stage and reverse it before it becomes clinically manifest. These approaches embody the first two Ps of P4 medicine: prediction and prevention.

Finding Cancer in the Blood

Cancer diagnostic biomarkers are molecules whose levels (or fluctuations) reflect a transition in health status. These can include proteins, metabolites, lipids, fragments of “cell-free DNA” released into the bloodstream when tumor cells die, and blood exosomes—small lipid vesicles, often pinched off from tumor cells, containing RNA, proteins, or other chemical clues as to the location of a disease somewhere in the body. Biomarkers can be found in blood, urine, tissues, and tumors, and they can be used for early cancer detection, to stratify patients into subgroups based on their likely response to disease and treatments, and to separate tumors into distinct subtypes for appropriate distinct treatments.

Two types of tissue yield the most biomarkers for cancer: the tumor tissue itself and blood. Specific panels of mRNAs and proteins in a tumor—even single cells—can help us classify the tumor and, in some cases, identify whether it is mild or malignant. To obtain biomarkers from a tumor you have to know where the malignant cells are and carry out biopsies. But that is not the case for blood. Most people don’t think of blood as a tissue, but it is really the most pervasive and interconnected tissue in the body, because it bathes all major organs and captures information from them in the form of proteins, metabolites, and signaling molecules. These are some of the analyte classes we discussed earlier in the context of our longitudinal phenome analyses of Arivale clients. Blood offers an invaluable window into health and disease for cancer and many other conditions.

At some level, this isn't really news. Blood has long been seen as important to oncology because it enables cancer cells to spread to distant regions. This is the primary means by which early-grade tumors, which are local and generally more treatable, become dreaded metastatic cancers that have spread throughout the body and are difficult to treat through conventional means. Blood test screenings for prostate-specific antigen (PSA) and carcinoembryonic antigen (CEA) have long been used by doctors to diagnose for prostate and ovarian cancers, respectively, although there is controversy associated with both.[21] But PSA and CEA are singular biomarkers that are only relatively effective in signaling cancer progression or recurrence. Often, it takes a panel of multiple biomarkers to see the relevant cancer signature in the blood.

One class of proteins that is of particular interest are organ-specific blood proteins. This approach is just now beginning to be applied to cancer diagnostics. Tissues are different in part because the cells they are made of are programmed by different genes, and thus they express different proteins at different levels of abundance. This makes it possible to identify proteins that are only, or mostly, made in a particular organ. When one is healthy, the levels are fairly stable over time. Quantitative changes in these organ-specific proteins in the blood can thus offer vital clues as to where in the body a disease signal is arising, and which biological networks have been disrupted by cancer. When it comes to cancer, longitudinal deep phenotyping with organ-specific blood proteins in large populations provides the possibility of seeing the signal and employing this signal to determine the origin of the disease.

The company Grail recently made a remarkable advance in cancer biomarker discovery. By analyzing the epigenetic profiles of white blood cells (the methylation patterns of cytosine, the C base of DNA, throughout the genome that modifies the expression of nearby genes), it was able to diagnosis about fifty different cancers.[22] To demonstrate the perceived power of this fascinating observation, Grail was recently purchased by the technology company Illumina for $6 billion, and their multicancer early detection test, called Galleri, was tested in a clinical trial and is now available.[23]

Let us now consider two separate approaches to the discovery of protein biomarkers for cancer.

Among the thousands of Arivale participants, we observed about thirty-five wellness-to-cancer transitions. Because we had banked blood samples collected prior to diagnosis, we could go back and measure additional analytes to look for potential signs that could be used to help future patients. Ten of these cancer patients had measurements taken at enough points in time that we could dive in and study the transition more carefully, including three individuals who were diagnosed with metastatic cancers. Research based on our longitudinal tracking revealed that in each of these three cases—but in none of the other Arivale participants tested—a protein known as CEACAM5 became elevated in the blood up to two years before clinical diagnosis.[24] This striking correlation isn't necessarily the cause of the metastases, but it does represent an intriguing biomarker candidate for early detection of cancers that might be at risk of going metastatic, particularly because it is known to be elevated in many metastatic cancer patients after clinical diagnosis. In each of the ten cancer transition cases we studied, we found biomarkers in blood drawn prior to clinical diagnosis that were significantly perturbed in their concentration levels when compared to normal. This suggests that it should be possible to predict the transition from wellness to disease years before the clinical manifestation of the cancer, a point we will return to below.

Blood biomarkers might also help physicians separate benign from malignant tumors. More than three million lung nodules are identified in patients in the United States each year, and while we've made great strides in imaging, biopsy, and bronchoscopy, these approaches are expensive, often invasive, technology-intensive practices that frequently fail to distinguish between benign and malignant nodules. If it were possible to make diagnoses easier and more effective, billions of dollars in unnecessary surgical procedures and complications could be avoided.

Many doctors are wary of data-intensive screenings, in part out of concern for what are known as false positives. What will they do if the blood work suggests that something might be amiss? Many women have had invasive breast surgeries that were probably unnecessary, and many men have had their prostates removed who could well have avoided that often debilitating surgery. Likewise, many surgeries have been performed to remove nodules from the lungs that turned out to be benign. One solution is to narrow the window of variance and to perform ancillary analyses aimed at weeding out false positives from among the blood biomarkers that are meant to distinguish between benign growths and malignant tumors. At ISB, we partnered with data scientist and bioinformatics expert Paul Kearney to use mass spectrometry, which can quantify hundreds of proteins from a very small blood sample to find a panel of blood proteins that could distinguish benign from malignant lung nodules. This capability was developed via a spin-out company that brought it to clinical practice.[25]

We used a systems approach involving multiple steps to identify this biomarker panel.[26] We needed to select a set of proteins that were relevant to lung cancer and from which we could identify a panel of proteins to make the benign / malignant lung nodule distinction. From the scientific literature, drawing on differential analyses of normal and cancer cell-surface and secreted proteins from lung cells, we identified just under 200 blood proteins that were possibly related to lung cancer and supposedly present in blood. We then quantified these proteins from the blood of seventy-two patients with diagnosed benign nodules and an equal number with diagnosed malignant tumors. About thirty-two proteins had some capacity to make this distinction—a promising set to continue our systems-driven analyses. From there, we created one million panels of ten randomly chosen proteins from the set of thirty-two potentially effective biomarkers.

In the most promising ten-member panels, we found thirteen proteins that were "cooperative," which is to say that they were found most frequently in malignant nodules. We found that twelve of these proteins mapped into three disease-perturbed networks associated with lung cancer—thus validating the relevance of these blood biomarkers. Clinical validation showed a negative predictive value of 94 percent. Later, this test was refined to a two-protein biomarker panel that was even more effective at making this distinction.[27] This relatively cheap biomarker panel can prevent unnecessary surgical procedures. It has been estimated that if this procedure were to be carried out across the United States, it would save the healthcare system as much as $3.5 billion a year by avoiding unnecessary lung surgeries performed in response to false positives.[28] This blood-based lung cancer screening technology is now being used in hospital systems as the Nodify Lung product from Biodesix. And how should we measure the human cost of this vital information for the people spared invasive surgeries and any complications that may arise from them?

We collaborated on a similar approach in partnership with Kearney and Jay Boniface, the chief scientific officer for the diagnostic testing company Sera Prognostics, to help them identify a two-marker panel for preterm birth—one of thousands of potential uses of biomarkers in a clinical setting.[29] This marker allows women to find out when they are eighteen to twenty weeks into pregnancy whether they are at high risk for preterm birth, permitting doctors to take additional steps to improve their chances of a healthy, full-term pregnancy. This test was developed into a product called PreTRM.[30]

Personalizing Treatment

Cancer is one of the fields that is increasingly moving away from a "one size fits all" approach to what is known as precision treatment. It has probably gone further than any other area of medicine toward personalization of treatments based on a patient's genetic profile. Genetic sequencing allows caregivers to know many of the mutations that make a cancerous cell different from the genome their patient was born with, and genetically informed therapies have been brought into common use through companies such as Tempus and Foundation Medicine, both of which sequence tumors to help guide treatment selection.

One of the earliest and most famous successes in this field was the drug Herceptin, which is used to treat breast cancers in women with a specific DNA molecular signal encoded by the genetic variant known as HER2/neu.[31] A striking association became clear in the early stages of clinical trials conducted by Roche-Genentech: women with the HER2/neu variant were having extraordinarily high rates of positive response to the therapy—far more so than those with other variants.[32] This prompted researchers to end the trial early, as it was deemed unethical to continue giving placebo controls to people whose cancer was likely to benefit from the real drug. Herceptin received early approval from the FDA in 1998.[33] The HER2/neu genetic variant was a "companion diagnostic" that identified the population of breast cancer patients who were most likely to respond to Herceptin. Trial results like this are becoming more commonplace, and companion diagnostics are being sought broadly not only for cancer drugs but also for many other conditions, matching treatments with specific biomarkers that indicate that successful treatment will be more likely.

Since that time, at least 140 companion diagnostics have been approved by the FDA for clinical use, with 42 percent of the drugs approved in the United States in 2018 and 25 percent in 2019 using a companion diagnostic.[34] Additionally, 65 percent of the drugs approved by the European Medicines Agency between 2015 and 2019 had at least one biomarker consideration in the drug development process.[35] As we get increasingly good at coupling the right therapies with the right biomarkers and apply deep phenotyping early on, we can greatly reduce the cost of drug discovery and target drugs to patients who can successfully respond.

Deep longitudinal phenotyping will help researchers solve fascinating riddles such as why 30 percent of individuals with myelodysplastic syndrome, a complex preleukemia known as MDS, will develop acute myelogenous leukemia while others will not. If a specific genetic variant, or combination of variants, is responsible for this transition, individuals without these variants will have less to worry about. Those who are at greater risk, meanwhile, can get aggressive treatment earlier in the process of disease transition, confident that it is a step statistically worth taking. Deep phenotyping may also help us better understand why, among individuals who develop acute promyelocytic leukemia, about 10 percent fail to respond to current combination treatment regimens. Identified early in the process of treatment, these individuals can immediately be moved to alternative therapies, saving critical time, cost, and, ultimately, lives.

Conducting genome/phenome analyses, or what is known as deep longitudinal phenotyping, enormously increases the amount of information we have on each individual patient, enabling us to understand their disease trajectory and to measure and track how the body's various systems are responding to the disease. Crucially, it also provides insights into how the relevant drugs work and which are likely to be most effective for any given patient. It makes it possible to enroll far fewer patients in clinical trials and will give us a chance to emerge with far deeper insights than can be achieved from conventional clinical trials. This will transform how we deal with cancers and so many other diseases.

Ending the Tragedy of Recurrence

If you've ever had a personal experience with cancer, you will know how fraught the word "cure" is. An initial treatment may seem successful. The tumor might be reduced to "undetectable" levels. Life may seem as if it has finally returned to normal. But often the cancer returns, sometimes in the same general area and sometimes at other sites. While the initial treatment may have killed off the cells it was intended to destroy, a small number built up a resistance to the treatment and survived. Worse, the disease is often more virulent the second time around.[36]

One of our recently launched studies at ISB, a $25 million effort funded by the state of Washington's Cancer Research Endowment (CARE Fund), aims to use genome/longitudinal deep phenotyping to monitor patients for the earliest signs of a cancer's return. Our hope is that this will help end the scourge of remission. Jim Heath is heading the project, along with Charles Drescher of the Swedish Cancer Institute in Seattle, who is leading the clinical efforts. The study includes a number of disease-specific measures, such as tumor genome sequencing, cell-free DNA, circulating cancer-specific exosomes (extracellular vesicles released from cells), and screening for additional cancer-specific proteins, as well as a deep dive into single-cell immune cell profiling. The goal of the CARE trial is to use this extensive phenotyping to identify colon, ovarian, and breast cancers at the earliest stages of recurrence and design highly targeted, personalized vaccines tailored to the specific cancer mutations of each individual patient.

Swedish Medical Center is the largest Providence St. Joseph Health hospital system in Seattle, and the head of the Swedish Cancer Institute was a member of our board at ISB (which is also affiliated with Providence St. Joseph Health). So they were natural and fantastic partners to collaborate with on this effort.

The scientific literature is full of descriptions of the effects of cancer and its treatments on quality of life, but the mechanisms determining these post-treatment outcomes are poorly understood. We launched the Surviving Through Breast Cancer trial in late 2019 with our partners at the Swedish Cancer Institute in an effort to fill in these gaps. The trial aims to help breast cancer survivors return to full health. In the United States, women have a one in eight chance of being diagnosed with breast cancer in their lifetimes, and every year, a quarter of a million women are diagnosed with invasive breast cancer.[37] Thankfully, this is one cancer in which tremendous progress has been made. There are now more than three million breast cancer survivors in the United States alone. However, as many will attest, health issues often linger long after the cancer has gone into remission. Some symptoms can be relatively mild, such as hair loss, though even these will impact survivors' long-term quality of life. Others, like cardiomyopathy—a condition leading to the degeneration of heart muscle cells—can be life-threatening. Breast cancer–associated disease and treatment can also induce symptoms such as severe fatigue, drastic weight change, cognitive impairment or "mental fog," depression, mouth sores, inflammation of the gut, and joint pain.

Where do these symptoms come from? In many cases, we don't know. We also don't know why they affect people so differently. Erin Ellis of the Swedish Cancer Institute and Andrew Magis, an ISB data scientist, believe it should be possible, through genome / longitudinal phenome trials, to better understand these variables. Ellis and Magis are collecting genomics data, clinical blood chemistries, blood metabolomes and proteomes, and gut microbiomes, as well as quality of life assessments, neuro-cognitive functional assessments, and lifestyle data. Once analyzed, our hope is that all this information will provide clues to the molecular underpinnings of many of the post-treatment negative symptoms of breast cancers—and point us in the direction of personalized interventions.

Toward a True Cure

Notwithstanding the tremendous progress being made in the detection and treatment of cancers, we do not have anything like a "cure." The slash-poison-burn approach successfully eliminates only a small fraction of cancers. Precision cancer therapies can be effective, but as of 2020, only 7 percent of patients with the right mutation profiles respond to this approach—although this figure is up from 2.7 percent in 2006, showing that we are moving in the right direction.[38] There are two further complicating factors: these therapies are very expensive, and they can leave patients more resistant to other treatments after patients have become resistant to their use. Immunotherapy holds enormous promise, but it cures only a few cancers (mostly tumors in the blood or liquid tumors, such as leukemias or lymphomas) and has not yet been effective for most organ or tissue tumors, or for those tumors with a lower frequency of mutations.[39] Clearly, we need other options.

Precision cancer treatment mostly targets cancers when they are quite far along, and we believe it's time to use the tools of scientific wellness to couple early detection with targeted and personalized treatment and to move from late-stage, high-tech, high-cost interventions to a strategy of "predict and prevent." We believe this can be done by carrying out genome/longitudinal phenome analyses of large, "real-world" populations to identify the earliest transition states—and then expand the range of safe and targeted therapies that could be used for very early personalized therapies. As noted earlier, among the 5,000 individuals we tracked at Arivale, thirty-five went from wellness to cancer over the four years during which we tracked their health. For the ten tumors we examined closely, we performed retrospective analyses to see if there were previously unrecognized signs in the protein biomarkers months or even years before the clinical diagnosis was made. Our hope is that this kind of intensive health tracking will make it possible for researchers to develop increasingly better early biomarkers so we can design therapies that will target the cancers at their simplest stage, before clinical diagnosis. Nathan is collaborating on President Biden's Cancer Moonshot 2.0 project, using 6,000 of Thorne's OneDraw at home blood collection devices to gather omics data for early detection strategies.

Lee is currently working to gain support for a ten-year, million-person genome/phenome project, which we will discuss in greater detail in Chapter 11. We estimate that such a project would likely identify more than 200,000 wellness-to-disease transitions for all types of chronic diseases, from cardiovascular disease and diabetes to cancers and neurodegenerative diseases such as Alzheimer's. We hope to provide statistically validated early-stage biomarkers for most chronic disease transitions, including many cancers, permitting us to attempt to design new therapies to reverse the disease transition before it is clinically detectable. This, we believe, is the most effective intermediate-term road to curing cancer.

One of the patients who signed up for the Arivale program would have been an ideal candidate for this "predict and prevent" approach. That patient, who we will call Diana, was in her fifties when she was diagnosed with Stage 4 pancreatic cancer—a grim diagnosis. She had been in excellent health before this, so it came as a complete surprise. Prior to her diagnosis, we had taken four blood samples, six months apart. We went back to the banked samples of her blood and did extra measurements to see if there might have been telltale signs in extended measurements that did not show up in the clinical lab tests. At that time, Arivale had around 2,000 clients, and we compared the levels of 1,000 or so individual blood analytes from Diana against the average levels of these analytes from the comparison group. We spotted a few clear outliers, far off the normal levels. We found five protein outliers in each of Diana's blood samples, three of which belonged to disease-perturbed biological networks known to be associated with pancreatic cancer. These differences appeared two years before the clinical diagnosis was made, suggesting that the earliest transition to disease may actually have occurred years earlier than her diagnosis and with no apparent symptoms.

It is important to note that these findings were done as retrospective research to learn from and build toward monitoring tests in the future. We unfortunately did not learn about these potential early biomarkers for this cancer until long after Diana had been given the devastating diagnosis of Stage 4 pancreatic cancer, one of the most aggressive forms of cancer with a typical survival time of only a few months after diagnosis and no known effective treatment. Frustratingly, even before Arivale closed, there was nothing we could do. Regulatory requirements in the United States do not allow communication of such information to an individual about outliers in these proteins without a prospective trial, FDA approval, and physician guidance. Our hope is to validate these kinds of early-stage markers, take them through clinical trials, and get to the stage where these signals can be used across the population at large to greatly reduce the incidence of cancers through prediction and prevention.

We were able to examine nine disease-perturbed biological networks for pancreatic cancer across Diana's four blood draws. In the earliest, none were disease-perturbed. In the second, one was perturbed. In the third, two were perturbed. And in the last, six were perturbed. This led to two important observations. First, the initial disease-perturbed network gave us clues as to where to search for a systems-driven therapy that may reverse this disease early. Second, we saw a striking increase in disease-perturbed networks as the cancer progressed. Our hope is that common cancers will share diagnostic markers and therapeutic strategies. With the million-person project, we will be able identify and validate early biomarkers and design effective early therapies for different cancers as well as other disease transitions. Then, with these powerful diagnostic biomarkers and appropriate therapies, we will be able to "predict and prevent" many chronic diseases.

The ultimate objective of the predict and prevent strategy is to carry out widespread genome/longitudinal phenome analyses so that everyone can be checked at each blood draw—ideally, across their lifetimes. How? First, we will have to establish the blood analyte norms for common cancers with, for instance, twice-a-year blood draws over extended periods of time. This would permit us to study each analyte from each blood draw for each individual and to compare it against the average population to identify potential outliers, giving us a personalized readout of what is unusual and a clear indication of how these markers have changed over time. Outliers would be potential wellness-to-disease signals that could help identify disease-perturbed network transitions much earlier, providing insights into the biological mechanisms of a particular cancer and possibly even suggesting therapeutic approaches. A functional analysis of these outliers over time would give us insights into individual patterns and fluctuations. The challenge, as always, will be recognizing when these fluctuations are just idiosyncrasies and when they could be a harbinger of disease. The data coming from the million-person project will be invaluable in this regard.

In the initial stages, we would target those at high risk for cancer to identify and track many more wellness-to-cancer transitions. These would include people with a family history of cancer and those with one or more of the twenty dominant genes that cause cancer (like BRCA2 for breast and ovarian cancers and Lynch syndrome for colon cancer) or with high cancer polygenic scores—the combination of factors that tell clinicians a patient is more likely to develop cancer. Another intriguing high-risk population is individuals who have survived cancer once already, who represent 20 percent of "newly occurring" cancers. For each of these high-risk populations, we will focus on prediction and prevention.

Once these initial studies have been carried out, clinical diagnoses in larger and broader groups will allow us to see which of the indicators from high-risk groups are also indicators in lower-risk populations.

In time, we will be able to predict the type of cancer directly from the perturbed analytes. The words "thankfully, we caught it early" should become the standard of care—uttered many years earlier than we could ever hope to do under current diagnostic paradigms.

This is the approach Lee presented to twenty-five cancer thought leaders, including researchers and oncologists, who began to work together at the invitation of Columbia University professor Azra Raza in June 2020. Raza's charge to this "oncology think tank" was to engage in an unbounded exploration of how we can move forward to cure cancer. Seventeen lectures were given by experts on this topic. When Lee gave the final lecture, offering our model for prediction and prevention focused on detecting the earliest cancer transitions and reversing them through a systems approach to therapy, including immunotherapy, there was widespread consensus that this was perhaps the most promising route for a cure. After the conference, the speakers, led by Raza, published a piece in *Scientific American* focusing on our "predict and prevent" approach.[40]

It is hardly surprising that oncologists would be enthusiastic at the thought of finding biomarkers in the blood that could lead to detection of cancer months or years before the first symptoms. But the predict and prevent approach could be generalized to virtually all chronic diseases. This would be a transformational moment in the history of medicine, as it could usher in the end of most chronic diseases for those who can access this new form of care.

Researchers don't have all the answers, but there is now significant momentum in the race to end cancer. Immunotherapy, new approaches to identifying blood biomarkers, and genome/phenome clinical trials will, we believe, move us aggressively toward effective cancer treatments in the next ten years. In 2020, the American Cancer Society reported the largest single drop in cancer mortality ever recorded.[41] While challenges abound, as they always will, the tide finally seems to be slowly turning.

10

The AI Imperative

Why Artificial Intelligence Is Essential to a Wellness-Centered Future

Given the accelerated pace of today's world, the conference that took place at Dartmouth College in the summer of 1956 might seem like an unimaginably unhurried affair. For two months, eleven scholars set aside their responsibilities to brainstorm, contemplate the future, and work together to catalyze the foundations of a new field.

The Dartmouth Summer Research Project on Artificial Intelligence was a gathering of founding fathers: Claude Shannon, whose information theory undergirds modern digital communications; John McCarthy, who coined the term "artificial intelligence" and developed one of its first computer languages, known as "LISP"; Marvin Minsky, cofounder of the first artificial intelligence lab at the Massachusetts Institute of Technology; and Herbert Simon, who would go on to win the Nobel Prize in economics and the Turing Award, the highest honor in computer science, for his contributions to psychology and decision-making. The contributions of these scientists, while enormous, were so oriented toward the future that even today their implications are just beginning to be realized.

We've shown you a vision for the future of healthcare that is fundamentally different from what many people have come to expect based on their experience with contemporary doctors and hospitals. The realization of this vision of a healthcare that is predictive, preventive, personalized, and participatory depends on many moving parts, but no tool will be more important than artificial intelligence. AI systems are already transforming healthcare. Those changes will accelerate in the coming years to such a degree that AI will soon be as much a part of our healthcare experience as doctors, nurses, waiting rooms, and pharmacies. In fact, it won't be long before AI has mostly replaced or redefined virtually all of these. As the dramatic expansion of telehealth during the COVID-19 pandemic has shown, when there is enough of a need, healthcare providers can pivot to adopt new strategies faster than we would imagine.

The Power of Processing Speed

There are two different, yet complementary, approaches to AI. The first camp takes the view that, given enough data and computing power, we can derive complex models to accomplish difficult tasks—a great many, or possibly even all, of the tasks humans are capable of. The data camp believes that all we need is data and lots of computer cycles to solve problems. Domain expertise in the relevant area is not required. Want to get a computer to drive a car? With enough data, you can do that. Need a robot to bake a cake? Data will get you there. Wish to see a painting in the style of Berthe Morisot materialize before your very eyes? Data and massive computing power can do it.

The second camp bets on *knowledge* and focuses on imitating how humans actually reason, using conceptuality, connection, and causality. The knowledge camp believes in the critical requirement of domain expertise, building algorithms to apply approximations of accumulated human knowledge in order to execute logic on a fact pattern via what are commonly called expert systems. These are often rule-based or probabilistic calculations, such as if a patient's HbA1c is higher than 6.5 percent and their fasting glucose is higher than 126 mg/dL, then there is a high likelihood that the patient has diabetes.

Today, data-driven AI is much further developed than knowledge-based AI, as the complexity of rules-based expert systems has been a significant impediment to scaling. The systems that enable self-driving cars to operate on our roads are all data-based. The algorithms that big tech companies use to guide ad placements, messaging, and recommendations are all data-based. As we will see, some important problems in biology are being solved brilliantly by data-driven AI, as well. But in an area as complex as human biology and disease, domain expertise may ultimately be more important in helping us make sense of the complex signal-to-noise issues that arise in big data. Indeed, it is likely that we will have to integrate the data-driven and knowledge-driven approaches to handle the extreme complexity of the human body.

Data are nothing without processing power. Neural network strategies have advanced enormously thanks to the demands of computer gaming, which provided the market forces that so often drive computational innovation. Gamers wanted realism and real-time responsiveness, and every advance toward these goals by one company stoked an arms race among others. It was in this hypercompetitive environment that *graphical processing units*, or GPUs, were developed to optimize the manipulation of images. If you've ever noticed how incredibly realistic video game characters and environments have become in recent years, you're marveling at the hyperfast renderings made possible by GPUs.

These specialized electronic circuits didn't stay in the realm of gaming for long. Andrew Ng, an AI leader and teacher of widely used online courses, was the first to recognize and exploit the power of GPUs to help neural networks bridge the gap between what the human brain evolved to do over millions of years and what computers have achieved over a matter of decades. He saw that the ultrafast matrix representations and manipulations made possible by GPUs were ideal for handling the hidden layers of input, processing, and output needed to create computer algorithms that could automatically improve themselves as they moved through the data. In other words, GPUs might help computers learn to learn.

This was a big step forward. By Ng's early estimates, GPUs could increase the speed of machine learning a hundredfold. Once this was coupled with fundamental advances in neural networks' algorithms, such as backpropagation, led by luminaries like cognitive psychologist Geoffrey Hinton, we arrived in the age of "deep learning."

What makes deep learning so deep? In the early days of artificial neural networks, the networks were shallow, often containing only a single "hidden layer" between the input data and the generated prediction. Now we have the ability to use artificial neural networks that are tens or even hundreds of layers deep, with each layer containing nonlinear functions. Combine enough of these and you can represent arbitrarily complex relationships among data. As the number of layers has increased, so too has the capacity of these networks to discern patterns and make predictions from high-dimensional data. Correlating and integrating these features has been a game changer.

Consider what we could do by applying that sorting power to an individual's personal data cloud. In goes the genome, phenome, digital measures of health, clinical data, and health status. Out come patterns recognized as indicative of early wellness-to-disease transitions and predictions of what choices might lie ahead with bifurcations in the disease trajectory (e.g., whether you could develop or avoid chronic kidney disease, or stave off advancing diabetes to regain metabolic health rather than progress to advanced stages with diabetic ulcers and foot amputations).

The potential is astonishing, but there are limitations to this approach. These high-quality predictions come from extremely complex functions, resulting in a "black box" that leads to a decision whose logic we can't fully comprehend. Deep nets are great "analogizers." They learn from what they see, but they can't tell you about something new. Data-driven AI can help us find functions that fit trends in data. It can work virtual miracles when it comes to statistical prediction, with nuanced and accurate predictive capability. But it can do no more than that. And this is a critical distinction. A world where we based our understanding and actions on data correlation alone would be a very strange world indeed.

How strange? Well, if you were to ask AI to tell you how to keep people from dying of chronic diseases, it is liable to tell you to murder the patient. Murder, after all, isn't a chronic disease, and if done early in life, it would be 100 percent effective at ensuring no death from chronic disease. The sorts of options that are so ridiculous or immoral as to be inconceivable for most humans are on the table for computers because ridiculousness and immorality are human concepts that are not programmed into computers. It takes human programmers—presumably those with decency and compassion and a sense of ethics—to write specific lines of code limiting AI's options. As Turing Award winner Judea Pearl put it in *The Book of Why*, "data are profoundly dumb."[1] Uberfast data are just profoundly dumb at light speed.

By "dumb," Pearl didn't mean "bad at what computers are supposed to do." Of course not. Computers are phenomenal at computing. What they're not so good at is anything else. Program a computer to play chess, and it can beat the greatest of human grand masters, but it won't have any way of deciding the best use of its power after the game is over. And it isn't aware that chess is a game or that it is playing a game.

This is something Garry Kasparov realized soon after his historic loss to IBM's Deep Blue. Yes, the machine had defeated the man, but Kasparov would later note that, from his perspective, it seemed that many AI enthusiasts were rather disappointed. After all, they had long expected computers to overpower human competition; that much was inevitable. But "Deep Blue was hardly what their predecessors had imagined decades earlier," Kasparov wrote. "Instead of a computer that thought and played chess like a human, with human creativity and intuition, they got one that played like a machine, systematically evaluating 200 million possible moves on the chess board per second and winning with brute number-crunching force."[2]

What happened next got far less press but was, to Kasparov, far more interesting. When he and other players didn't compete with machines but instead teamed up with them, the human-plus-computer combination generally proved superior to the computer alone, chiefly because this melding of the minds changed their relationship to perceived risk. With the benefits of a computer able to run millions of permutations to prevent making a ruinous move or missing something obvious, human players could be freer to explore and engage in novel strategies, making them more creative and unpredictable in their play. This might not always be the case when it comes to games, which are closed systems where brute force and number-crunching ability are incredibly powerful, but we believe it is a vital lesson for twenty-first-century medicine, because, ultimately, when it comes to health, it is not enough to spot patterns: we need to understand biological mechanisms and to know why things happen as they do so that we can intervene appropriately.

The future of healthcare will take us to a place where increasing numbers of routine medical decisions are being made by AI alone. But far more decisions will come from a combined approach of powerful AI assessments augmented and amplified by highly trained human intelligence, a schema that has come to be known as "centaur AI." Like the mythical half-human, half-horse creature of Greek mythology, this hybrid arrangement is part human, part computer and should offer us the best of both worlds. This is especially true in areas where extreme human complexities play major roles and brute computational power is likely to be less successful than it can be in a closed, fully specified system like a game.

Science's 2021 Breakthrough of the Year

There is a long-standing "grand challenge" for computational biology: being able to predict the shape of a folded protein given just its gene (amino acid) sequence. This is an ideal problem for data-driven AI in that it is well defined, offers concrete ways of measuring "better" or "worse" predictions, and can draw on large data repositories, such as the Protein Data Bank which has the three-dimensional structures of thousands of proteins for training. It's also a question of great importance, because the shape of a protein determines its function—what chemical reactions it may catalyze, what molecular machines it will help build as multi-protein complexes, and how it interacts with other molecules to assemble cells, tissues, and organs. So if we want to understand the chain of events that moves us from DNA to RNA to amino acids—the building blocks of proteins—and onward to the full complexity of human life, we have to know how proteins fold.

This challenge of predicting protein structures based on their gene (amino acid) sequences—and trying to get these predictions as close as possible to actual experimental measurements—is so important to the scientific community that a competition is held every two years pitting the best minds in the field against one another in an effort to stimulate a breakthrough in predicted folding accuracy. This competition, known as Critical Assessment of Protein Structure Prediction, or CASP, has been going on since 1994, seven years *before* the completion of the Human Genome Project. Brilliant researchers such as David Baker at the University of Washington and Yang Zhang at the University of Michigan have consistently won this contest and yet, for a period of eight years between 2008 and 2016, accuracy ceased to improve. During that time, the top performance at CASP hadn't changed much, with winning performances at around 40 percent on the Global Distance Test, or GDT, which is used to measure the difference between predictions and the experimental measurement of test proteins.

That changed when the Google-owned AI company DeepMind entered the fray for the first time in 2019. DeepMind had become famous for its work on AlphaGo, a program that defeated world champion Lee

Sedol in 2016 in the ancient game of Go, widely regarded as the most complex board game in the world, though its rules are quite straightforward. One player has a set of black pieces, the other white. The players take turns placing pieces on a grid, anywhere they choose. The goal is to surround the pieces of the other player. Hidden in this seeming simplicity is deep complexity: the combinations of possible moves are greater than the number of molecules in the known universe. Strategy is devilishly complicated, such that large-scale number crunching alone is insufficient to win. To make AlphaGo a success, its programmers had to leverage the 3,000-year history of human experience playing go. They then used reinforcement learning to iteratively improve the program through millions of games. Sedol, the eighteen-time world champion who fell to AlphaGo, would later say that because AI would rapidly improve—while humans are only capable of improving a little—it was unlikely that any human would ever again be as good at Go as a capable computer.

Following on the success of AlphaGo, the team at DeepMind embarked on an even more ambitious algorithm. They threw away everything that had been learned by humans about go, seeded the reinforcement learning algorithm with the rules of the game, and had the algorithm play against itself over and over. The modifications of the winning side were preferentially kept while those of the losing side were eliminated. The algorithm grew stronger each time it played the game. The resulting program, AlphaGo Zero ("Zero" to denote no human knowledge contamination) beat the original AlphaGo in a tournament one hundred games to zero. It was a stunning repudiation of what we thought we had learned playing go as humans. Expert players are often baffled, unable to understand why AlphaGo makes the moves it does, but that they are superior is not in doubt.

Could such a technology help us solve the protein folding problem? When DeepMind's AlphaFold entered the CASP contest in 2018, it beat the competition by nearly 50 percent, coming in at an accuracy of nearly 60 percent GDT.[3] It was an astonishing leap forward. In 2020, AlphaFold 2.0 blew the original AlphaFold away, making an even bigger leap forward and jumping to nearly 85 percent GDT.

The conventional "gold standard" for determining a protein's structure is made by forming a regular array of individual protein molecules stabilized by crystal contacts, a process known as protein crystallization.

This highly ordered form makes it possible to use techniques such as X-ray crystallography and nuclear magnetic resonance to define the protein structure. But this isn't the exact shape that the protein has in the body as part of the living system, where function can influence form. Researchers estimate that they can experimentally measure these structures with about 90 percent accuracy. That's just about where AlphaFold was back in 2020, meaning that we are now living in a world in which our computational predictions can be just as good as our experimental measurements for protein folding. Importantly, computational predictions at this level of accuracy could better represent what the structures are in their natural state (rather than in a crystalized state), making them even better than experimentally determined structures. A remarkable advance!

Such computational predictive power provides tantalizing opportunities, allowing us to simulate conditions we can't simply measure. But how do we know if the predictions are correct if we can't check them against direct measurements? One approach is to simulate effects that can be seen, and then determine whether those ancillary downstream consequences match. Will simulations of cell function be more or less accurate when using protein structure inputs from AlphaFold predictions or from experimental data? At this point we can't know—at least not completely. To some degree, as they say in Great Britain, "the proof is in the pudding." If computational predictions ultimately prove better than direct measurement at identifying underlying states of wellness, transition, and disease—and do so again and again across various challenges—it will be hard not to trust those predictions. Will they be perfect? Probably not. Will they have a better degree of accuracy than most human doctors achieve? In time, undoubtedly yes.

Programs like AlphaFold are computationally expensive. It can take weeks on a present-day supercomputer to simulate a complex protein. But when a team led by David Baker at the University of Washington took what they had learned from AlphaFold and integrated it with insights from their own work, they developed a hybrid human-computer algorithm, RoseTTAFold, that came very close to the accuracy of AlphaFold 2.0 but completed the computations in a fraction of the time, taking only about ten minutes on a single high-end GPU machine.[4] The time will continue to decrease as computing power continues to increase, so we will soon have quick, high-accuracy solutions to replicating the results of protein crystallization experiments that combine purely computational approaches with domain expertise in protein folding. This new ability to predict protein structures exemplified by AlphaFold and RoseTTAFold was *Science* magazine's Breakthrough of the Year for 2021—and rightfully so. They were revolutionary.

None of this suggests that the protein folding problem is completely solved. Far from it. Proteins interact dynamically and have multiple configurations as they carry out their functions in living systems. And while AI is now good at predicting the crystallized structure, none of the current approaches can predict proteins in all of their potential structures, and validating such predictions is a significant challenge. There is an entire field of molecular dynamics simulation devoted to this problem that uses the crystallized structures as a starting point, but these simulations are computationally demanding, even over very short periods of time.[5] It's almost undebatable, however, that AI and big data have reached a tremendously important milestone, one that is vital to increasing our understanding of scientific wellness.

Knowledge Is Power

For many challenges, data-driven AI is king. In the long run, however, it will take the power of both data-driven and knowledge-driven AI to fundamentally change healthcare. Quite logically, this will begin with the data-driven systems that are most advanced. And among the modern marvels of data-driven AI are artificial neural networks, inspired by the wiring of the human brain.

For computers to become faster, more efficient, and better at problem-solving, it made sense to model their functioning after the circuitry of their creators. While neural networks have been around for a long time—they were first proposed in 1949 by Donald Hebb, who gave us the "what fires together wires together" conception of neural learning discussed earlier in this book—they weren't very effective until recently due to limitations in data available and computational power.[6]

We have much more data at our disposal today. Enormous caches of electronic health records, or EHRs, have been amassed by following patient interactions with healthcare providers across essentially all major health organizations. A 2017 study found that an average of 80 megabytes of information were generated by each patient per year.[7] These records include imaging data, basic testing results, information on patient outcomes, and more.

One of the tasks doctors understandably seem to hate more than just about anything is inputting patient data into EHRs at the end of a long day. This has given rise to a new, dark joke among physicians, who often lament that they end each day by making a sacrifice to their robot overlords. All of this comes before we add in the larger data sets that will become part of each patient's data cloud when we start amassing and integrating genomics, longitudinal phenotyping, gut microbiome analyses, and data from wearable devices. But these will automatically be fed into the medical record and won't have to be manually and laboriously logged. (Lee went to his fiftieth medical school reunion recently and found that many of his colleagues had retired; the strongest motivation for retirement, for most, was detailing their patients' EHRs and dealing with billing.)

Another major domain of medical AI is interpreting imaging data. Images make up a large fraction of medical data and generally take significant amounts of time from trained experts to interpret.[8] AI technologies are now providing help in extracting, visualizing, and interpreting imaging data, in some cases generating insights that extend beyond what humans are able to do. Deep learning is at the heart of many of these algorithms.[9] Companies such as IBM's Watson Health, Google's DeepMind, Microsoft's Open Mind, and others are building capabilities for many important applications, including detecting anemia and identifying various cancers.[10]

A team at Google was able to use deep learning to identify blood vessel damage in the eye with very high sensitivity and specificity after training on 130,000 retinal images.[11] Importantly, the diagnostic performance of this algorithm was essentially the same as the results achieved by US board-certified ophthalmologists. In a study conducted at Stanford University, AI was able to detect arrhythmias on an electrocardiogram with higher accuracy than the average cardiologist.[12] For now, we should think of these systems as aids to clinicians, but it is not hard to image that we will eventually be entering a world in which the analysis of medical imaging data will mostly be handled by computers, making it possible to incorporate much larger imaging data sets and process them more quickly to readily provide information for decision making. Because AI is generally inexpensive to run once it has been developed, the potential for optimizing care and making it radically cheaper is striking.

The Use of Knowledge-Based Systems

We are discovering new ways to code collective human knowledge into computers, a class of AI approaches built on what have long been called "expert systems." In a perfect world, these systems would be able to execute decisions on a set of facts and come to the same conclusions as human experts—or even improve on human performances with lightning-fast processing speed, perfect memory recall, and the ability to see the nearly limitless permutations that could arise from any given data combination. At their best, they are akin to having not one expert but thousands upon thousands, all working together at top speed. That is the goal, and we are well on the road to achieving it.

Traditional expert systems have been hard to scale because they tend to get convoluted as the rules pile up, leading to incredibly complex decision trees. Also, human thinking is not purely rule based. Humans are quite good at recognizing when rules shouldn't be applied to a particular case or where the logic breaks down. While the more common cases can effectively be captured by AI, enumerating every possible permutation for a computer is an impossible task. One of the marvels of the healthy human brain is that it does not get stuck in endless loops and is broadly able to deal with the unexpected. A breakthrough similar to the one that propelled data-driven AI systems is sorely needed for knowledge-based AI systems. This would lead to a world in which "deep learning" could be joined by "deep reasoning," such that AI can understand implicit relationships, not just ones that have been specifically programmed into its code. What makes this challenge so difficult is that unlike deep learning, where adding massive amounts of computing power and data fueled the leap forward, we need conceptual advances to make deep reasoning achievable.

Before we can hope to get there, we need to understand what reasoning is. One way to get to this is through a seemingly unrelated question: Do platypuses drink water?

Well, you might think, of course. You don't need to be informed by massive amounts of data on platypuses and water to come to this conclusion, as machine learning would require. After all, you likely know that a platypus is a mammal—albeit a very strange sort of mammal—and you probably think it's fair to assume that all mammals need water. So the chances are good that platypuses drink water. But unless you happen to be a zookeeper or a platypus expert, how would you know for sure? Have you ever seen a platypus drinking water? Is the fact that platypuses drink water written down in anything you've ever read from a credible source?

Just to be sure, you'd probably do a quick Google search of the question "Do platypuses drink water?" And do you know what you would find? No specific answer. What's more, platypuses spend a lot of their lives in the water, so if your idea was to jump onto a zoo's webcam and look for a water bowl, you'd be out of luck. If they are drinking water, they're likely doing it while swimming.

But if you had to hazard a guess, you'd probably go with your intuition, relying on implicit logic to guide your thinking, because that's how humans make decisions when they don't have perfect information. Data-driven AI, though, has a difficult time with such questions because it has no intuitive understanding of mechanism or causality. It doesn't do well at "guessing." It lives in a world of correlation and prediction.

A form of AI that brings together today's incredibly powerful data-driven advances strengthened by a breakthrough in causal knowledge models will be much more comfortable with unknowns, implicit relationships, and implied probabilities—the sorts of things experts rely on to make decisions every day when they apply conceptual knowledge (domain expertise) to questions that are far less random and far more consequential.

AI Tools to Help Physicians

A host of AI tools have already emerged to help physicians with their diagnoses. Just a few years ago, most medical decisions were based entirely on the knowledge in the head of the doctor at the time the decision was made, though it has long been clear that the data deluge coming from biomedical sciences exceeds any human's capacity to process it even superficially. Today, clinical decision support systems have arisen to present physicians and other care providers with access to a wealth of information at the point of care. This leverages what computers are naturally good at—storing, recalling, and correlating vast amounts of information virtually instantaneously—and links it to an expert human's deep ability to reason intuitively and think creatively.

When these expert systems first came along in the 1980s and 1990s, they were met with hostility by many physicians who worried that computers would soon be in charge of medical decision making, taking the "doctor's touch" out of the equation and binding the hands of physicians whose opinions differed from the computer's analysis. But that's not what happened. Research has shown that these systems have gotten better and better at helping doctors spot potential outcomes they might have missed without taking the ultimate decision-making authority out of their hands.[13] The physician can still say no—at least for now.

We are fast approaching a time when "centaur doctors" combining the best parts of human intelligence and AI assistance will be empowered to make bold medical decisions with far fewer unintended consequences. That's vitally important, because medical mistakes account for about a quarter of a million deaths annually in the United States alone.[14] (Setting aside the recent COVID-19 pandemic, these errors are the third leading cause of death in the nation after heart disease and cancer.)[15] It is not even a little bit bold to say that AI-enabled healthcare already has saved countless lives.[16]

An AI program called MedAware has helped doctors avoid accidentally prescribing the wrong medication.[17] The system was pioneered by Dr. Gidi Stein after he heard about a 9-year-old boy who died because a doctor clicked the wrong box, ordering up a prescription for blood thinners instead of asthma medicine. Mistakes like this are frighteningly common. About 70 percent of medication errors that may result in adverse effects are prescription errors.[18] And it's not hard to understand how this could be such a pervasive problem. The FDA has approved tens of thousands of prescription drug products, many of which have very similar names. There's Novolin and Novolog. There's vinblastine and vincristine. There's hydroxyzine and hydralazine. If you recall that doctors have famously bad handwriting, you can imagine how this could have been problematic in the days in which most "scripts" were written by hand, but it's still a challenge in the digital age, when a simple misspelling or temporary lapse in memory can deliver the wrong medication to a patient. So when a doctor prescribes a medicine that doesn't match the patient's medical needs as assessed by MedAware, that physician gets an alert. The system also signals doctors if they attempt to prescribe a medication that could interact negatively with one of the patient's existing medications—another common error that is almost never checked by physicians.

In hospitals around the world using MedAware, the doctor still has the final say. Sometimes an unusual prescription is warranted in a particular case. The system simply offers an extra check—one that is particularly beneficial when physicians are overworked and exhausted. And it is saving lives.[19]

There's another advantage: the risk of making mistakes often keeps doctors from thinking creatively, restricting their options to a small number of familiar treatments. These practices, at their best, are grounded in clinical trials, but with the combined power of AI and individual data clouds, we can do much better than "following the average," taking an individual's unique genetic makeup, biochemistry, lifestyle, and personal history into account. By eliminating simple errors and making available a wealth of scientifically validated insights specific to each person, an AI-assisted doctor could quickly and confidently evaluate tens of thousands of possible outcomes—in the context of each patient's unique biology and medical conditions—before settling on a much smaller selection of high-quality recommendations.

We're Not Quite Ready for Robot Overlords

Today, even in hospitals equipped with the best clinical decision support systems in the world, we run all key medical decisions through a physician. In our view, that shouldn't change in the short run. By virtue of their medical training, doctors have a wealth of knowledge, experience, wisdom, and judgment. Yet the human brain—even the greatest of human brains—cannot absorb, remember, or interpret even a tiny fraction of the information now available on human health and disease or encoded in our personal genomes and our molecular phenotypes. In an age of exponentially growing information, the expectation that everything should be squeezed through the cognitive limitations of the human brain is increasingly problematic. Unlike humans, AI systems can store and rapidly execute commands based on vast amounts of information. For tasks that can be narrowly defined and performed, AI can far outperform its human counterparts.

To a great extent, we already acknowledge this in our actions. When you search a list of symptoms online, find a match, and purchase a nonprescription medication, you're letting a searchable database help you make decisions without a doctor. This doesn't come without risks—including misdiagnoses, the sometimes dangerous use of unregulated supplements, and vulnerability to "cyberchondria," in which one comes to be convinced one has a disease after reading about symptoms on the internet.[20] Trusted sources such as WebMD, the Mayo Clinic, or the Cleveland Clinic can help us find accurate information online, but clearly there is a depth of expertise that working directly with a clinical professional can provide. Most people don't understand science well enough to reliably diagnose themselves. Nonetheless, people are increasingly trusting their health decisions to algorithms and artificial intelligence.[21]

How far AI can go to reduce the day-to-day decisions doctors make is an open question. Still, clinical decision support systems are now widely used, which represents a paradigm shift in healthcare.[22] They are commonly integrated with electronic health records or other software systems so they can be used at the point of care to provide ready-to-hand information for a clinician to inform their on-the-spot decision making. The big advantage is their capacity to mine data on a scale that would be uninterpretable by humans.

These systems are used today for prescriptions, diagnostics, managing disease treatment, and generating documentation to improve clinical workflows.[23] General areas in which clinical decision support systems aid healthcare professionals today can be bucketed into different groups, including patient safety, clinical management, cost containment, diagnostics support, and patient-facing decision support.[24] One clinical decision support system alerts physicians when a patient's electronic health record shows they are eligible for a clinical trial.[25] Another helps ensure documentation accuracy (to make sure, for example, that a patient is vaccinated following surgical spleen removal to reduce the risk of infections in surgery, including infectious agents such as *Haemophilus influenzae*, pneumococcus, and meningococcus).[26]

These systems are getting better and better. One system, used for the diagnosis of peripheral neuropathy, has achieved an accuracy of 93 percent compared to experts at identifying motor, sensory, mixed neuropathies, or normal cases.[27] It's easy to say, "But that's not as good as a real doctor!" But not every patient has access to a real doctor. Systems such as these are tremendously useful in places with limited entrée to clinical experts, and they can help bring a higher level of specialized care and diagnostic precision to patients who would not otherwise find high-level care easily available.

While there is still some resistance, "real" doctors are increasingly clamoring for AI systems that can be used as aids. To give one example, the DXplain tool, a decision support system developed at the Laboratory of Computer Science at Massachusetts General Hospital, provides a probable diagnosis based on clinical manifestations and furnishes the computed diagnostic and evidence used by clinicians. A randomized control trial involving eighty-seven family medicine residents showed that using DXplain resulted in significantly higher accuracy rates (84 percent versus 74 percent) on a validated diagnosis test involving thirty clinical cases.[28]

Clinical decision support systems are also used these days to help with laboratory testing and interpretation, providing alerts and highlighting abnormal lab results. They sometimes help avoid the use of riskier or more invasive diagnostic strategies in favor of safer alternatives. Liver biopsies, for instance, are considered the gold standard for diagnosing hepatitis B and C, as noninvasive lab tests are not considered accurate enough to be accepted. But AI models can combine multiple data sources, including imaging, blood markers, and genetics, to gain much higher accuracy rates without the need for a biopsy.[29] Clinical decision support systems also help make test results more personalized, adjusting the ranges for age, sex, ethnicity, disease subtypes, and so forth.[30]

Pathology reports drive many critical medical decisions, and AI can be used to perform tasks such as automated tumor grading. A study showed that urinary bladder tumor analysis could be performed with AI at an accuracy rate of 93 percent; about the same rate of accuracy has been shown for brain tumor grading and classification.[31]

Some doctors will fight against this; that's what happens when diagnostic power balances are in flux. But those who fold these systems into their practices will be doing their patients (and themselves) a great service. Using knowledge models, doctors can better ensure they're not missing a diagnosis or treatment that they either considered and dismissed or hadn't considered in light of their patient's unique biogenetic attributes. By feeding genomic and phenomic data from individual patients into a deep knowledge system for analysis and interpretation, doctors can derive leading-edge diagnostic and therapeutic approaches uniquely targeted to the individual.

What doctors need are systems that offer up actionable insights in ways that are transparent, interpretable, transportable, and open to iterative improvement. To that end, a key step is the creation of "logic models," computer-generated representations of logical processes in response to personal data, conditioned on large "knowledge graphs" that represent vast interconnections and different kinds of known relationships among presenting pathologies.

Executable knowledge models have paradigm-shifting potential. Whether knowledge is communicated orally or through written instruction, or learned by experience, there is always a loss in detail and interpretation. With knowledge models, transmission can be delivered exactly and executed appropriately on the other end. Such a capability, represented in part today through a variety of clinical decision support systems, can raise the available knowledge level of any physician closer to the level of the domain expert. In theory, at least, this makes it possible to bring the thinking of the world's best experts to any doctor, anywhere, for any given topic area. But even in the United States and Europe, where such systems exist, relatively few doctors know about them or are prepared to avail themselves of their use. Medical schools must create courses that bring an awareness of these transformational AI opportunities to young physicians. The million person project with genome / phenome analyses over ten years that Lee is proposing will generate thousands of new actionable possibilities—and these will be delivered to physicians using AI to clearly explain how to execute the possibility along with the logic of its rationale.

It's key, at least for now, that this knowledge be presented in a "clear box" manner, so that each step in the diagnostic or therapeutic reasoning chain can be reviewed and evaluated by the physician in charge. Such systems can create "supercentaur" doctors who are freed for higher-level thinking and contextualization. When exceptions are made and those exceptions are successful, the logic that drove the decision can be rapidly incorporated into the knowledge model. And because these knowledge models are fully transmittable, after validation, new knowledge can spread instantaneously—a far cry from the painfully slow process of the individual-to-individual education and dissemination practices of today.

The Promise of AI-Human "Relationships"

We are on track to meet a future in which computer systems can reason, decide, and explain their decisions to humans for final approval. Indeed, we believe such "centaur" systems will soon become essential in order for doctors to function in a world of exploding data and medical understanding and insights. In the versions of this collaboration most likely to be realized in the immediate future, AI will offer a selection of proposed decisions explained in terms that human physicians can understand, giving physicians a chance to critically assess the underlying reasoning. This closely matches how a medical specialist might guide a family practitioner to make the ultimate care recommendation to a patient. AI will become a powerful teaching tool.

Eventually, with advances in natural language processing, humans and computers will be able to engage in sophisticated, reasoned discussions about a patient, thinking through possibilities together in a real-time collaborative exchange. The doctor will be the final arbiter of what gets decided while being relieved of the impossible burden of learning and integrating the troves of new data and knowledge required to make the most informed decision. Deep reasoning will in time begin to identify connections and concepts that humans simply cannot see or understand without the help of AI. When this happens—when AI begins surfacing complex insights rather than just correlating data—what will doctors do? If these interventions work, will they be incorporated into the knowledge model even if we humans cannot explain why they work? That might sound like a step too far, but it's important to understand that we know that many pharmaceutical remedies work without completely understanding the biochemical processes at play. In fact, it's always been this way. The ancient Sumerians and Egyptians were creating medicines from salicylate-rich plants such as willows long before humans understood why such chemicals reduce pain and inflammation. And even though the modern drug derived from these practices, aspirin, has been around for more than 165 years, the mechanisms underlying its effect on the human body are still being studied today.

So how much trust should doctors place in AI? This is a vitally important question. The time is fast approaching when doctors will more regularly face the choice between following AI advice they may not fully understand, even with clear box explanations, or rejecting insights that might help prevent or ameliorate disease or avert death. Venturesome doctors will use these AI possibilities and, as increased experience with AI and patient outcomes accumulates, the answers should become clear. In the end, we suspect AI will be like aspirin—we will use it because it works, even if we don't fully understand it. It will take a leap of faith, but innovation often involves venturing into the unknown for a significant time before the unknown becomes "known."

Moving Forward with Digital Twins

One of the most exciting technologies under development today enables us to simulate each individual's unique physiology (top down) and biochemistry (bottom up) in a computer, creating a "digital twin" of the sort we described in Chapter 8. These programs essentially seek to create a computational version of all that is known about your unique physiology and biochemistry so that, when the time comes, medical interventions intended to help save or extend your life could first be simulated on your computational twin. If the biodigital twin responds in a healthy way to a treatment or therapy, it is likely you would, too. And if it has an adverse reaction, you would be warned away from this medical approach.

The usefulness of a digital twin should be immediately clear. "Let's try this and see what happens"—a philosophy that implicitly guides so much of our prescription practice—is not a winning healthcare strategy. A digital twin can be restarted, rebooted, and improved over time. It is a personal crash test dummy—a predictive model of your unique biology that integrates all your health and medical data to provide an on-demand resource for assessment and predictive decision support for patients and their care providers.

With the right data and simulation power, the digital twin can be used to test the likely outcome of different treatment choices—or the consequences of avoiding treatment—aiding in selecting interventions that are most likely to work with a higher frequency of success. It would help your physician determine whether your biochemistry and physiology are suitable for the use of a new therapy to stay healthy, fend off transitions, or fight disease.

The most effective digital twins are likely to be based on holistic physiological modeling, which aims to maintain the underlying balance of human physiology and biochemistry, or homeostasis—the essence of systems medicine. Such an approach provides a framework for understanding the continuum between health and disease, shifting the focus from a linear "falling dominoes" view of disease to a perspective that sees disease as a multidimensional imbalance in the normal processes that maintain our lives and health. This shift counters the blind spots that arise from approaches that look for one "dominant" mechanism and a single drug to intervene that then gets applied to everyone with the same diagnostic label. The widespread use of digital twins, unique for each of us, will allow us to become much more strategic at targeting therapies for specific individuals and will open the door to a much greater use of multimodal therapies.[32]

The true strength of the digital twin lies in its ability to reveal, especially for chronic diseases, effective multimodal interventions tailored to individual patients. We can use what we know about a person's unique biology to simulate the likely effects of interventions in ways that would be impossible to play out in real life, where each intervention takes time to evaluate and the price of failure is too high to mix and match. A digital twin's response to interventions takes place on a timescale that can be sped up such that years can be assessed in seconds. As a result, individual comparisons can be run against millions upon millions of hypothetical (or, in time, real) digital twins and assessed for all manner of comparative long-term results, with the ultimate goal of finding a set of actions and interventions that retains a homeostatic balance in the most effective way over the duration of a person's life.

We can't build these programs without an integrated and mechanistic view of each individual's body. For a twin to work, relevant phenomic and personal data must be frequently uploaded to it. Your genome is essential, as it defines your unique biology and impacts many critical health decisions. While much of the phenomic data relevant to health are expansive, they can be abstracted to help preserve privacy. Not every meal matters, but your general pattern of eating does. The day and location of your workout doesn't matter, but the kind of exercise and frequency with which you do it over the long term really do.

What's more, the assessment of each patient's situation is predicated on an ability to work with incomplete or uncertain data, because even though such a simulation requires a significant amount of data, there's always going to be something missing. Not everyone will have access to the same amount of information, and measurements may be made on different platforms. Some data that might be relevant won't be easy to collect, and some you may not want to share. Some patients may not be able to afford the time or cost of longitudinal or, for digital health, continuous monitoring or have the discipline to maintain regular blood draws, scans, evaluations, and digital health information. In these cases, a digital twin can be formed using de-identified data for other patients whose profiles are as similar to the patient as possible. It is important to remember that, while we are all unique, we have a lot in common. An improvement in knowledge for any one of us is thus an improvement for all.

Digital twins are crucial tools for N = 1 medicine, helping physicians treat every one of their patients more effectively, whether those patients play an active role in monitoring and assessing their health or not.[33] Digital twin and AI capabilities will also offer significantly increased efficiency and lower costs overall in providing personalized care. For example, we can target drugs or other interventions better to the small fraction of patients who are responders and better individualize prevention to avoid costly, prolonged, late-stage disease with negative health outcomes.[34] Such gains have the potential to lower costs for patients and benefit providers and insurers as well. Because of this, realistic economic models will need to convince private insurance or government plans to play an important role in making these advances broadly available to anyone who wants to opt in.

But the creation of digital twins, like health AI in general, comes with important challenges related to unequal access, inconsistent monitoring, and especially privacy, for it necessarily involves amassing a lot of detailed health information about your body in order to guide treatments and design a health optimization plan that is ideal for you.[35] This includes your genome, multiomic molecular measurements from the blood, gut microbiome, data from wearables, health history, clinical conditions, dietary habits, and environmental exposures. Just consider what those with nefarious purposes could do with all that information. It's critical that clear rules that govern the control of these data and what they can be used for are in place.

All medical information is strictly regulated in the United States, with legal guidelines for its use and dissemination under HIPAA, the Health Insurance Portability and Accountability Act, passed in 1996. Healthcare systems and all HIPAA-compliant organizations are required to have safeguards in place to control access and use of medical data, which are also regulated by consent forms from the patient. Still, more can and should be done to better balance the sometimes competing values of efficacy, affordability, discovery, equity, and privacy.

One early but potentially very robust class of solutions for the future comes out of advances in shared databases known as blockchains, which are most popularly known as the technologies behind cryptocurrencies but might actually be a better fit in healthcare. The decentralized nature of blockchain architecture makes it difficult for anyone to surreptitiously and nefariously change the code to gain access to personal information and assures significant transparency as to where data have gone and who is using those data. These are big steps toward the democratization of data sharing, putting individuals in complete control of who may access their data and for what purpose. As these systems advance, we expect they will become an integral part of healthcare's future.

The bottom line to all of this is trust. If the rules that govern privacy don't ensure broad confidence, the future we envision will assuredly collapse. But AI is fundamental to a future of healthcare in which medicine is safer, more effective, less intrusive, and less rife with suffering, so we must get this part right.

Our digital world will increasingly be able to provide more of our environmental exposure data—and especially the relevant parts that influence our health, such as smoke exposure from forest fires. Digital twins are crucial tools of N = 1 medicine—teaching physicians how to treat every one of their patients more effectively and uniquely, whether their patients play an active role in monitoring and assessing their health or not.

AI will be critical to delivering actionable possibilities to doctors in a concise, compelling, and clear manner. It will help ensure that the right diagnoses and treatments are provided to the right patients at the right time. And the right time, if we really want to improve lives, is early enough to stop disease before it starts. But before we can do this, we need to learn more about what the proteins coursing through our blood are trying to tell us. And that, as is often the case in the world of AI, will take a lot of data—or, to put this in human terms, a lot of individuals' medical life stories.

11

The Path Forward

What It Will Take to Ensure That the Future of Healthcare Arrives Sooner Rather Than Later

Our vision of what twenty-first-century healthcare can be—what it *should* be—begins with four words: predictive, preventive, personalized, and participatory.

Perhaps we do ourselves a disservice by whittling this ambitious new paradigm down so much, but the truth is, despite all the technology and innovation involved in bringing us to this point, it is really not all that complicated. We simply need to follow what the science permits and distance ourselves from an approach to medical care that has us managing illness instead of keeping people well. You don't have to look long or hard to know that the disease-oriented approach to healthcare isn't working. It doesn't catch disease soon enough, seldom prevents it from progressing, rarely treats patients as unique individuals, and almost always fails to inspire people to be active participants in their own journey to better health.

It doesn't have to be this way. By assessing every person's health with modern capabilities—taking many more measurements and using data-driven strategies to determine individual needs—we can optimize individual well-being, identify personal risks, extend health spans, and take steps to prevent or reverse the onset of disease. We can ensure, with a far higher degree of accuracy than is now the norm, that the treatments doctors recommend are right for each person, and we can give individuals the tools and support they need to take charge of their own health. Doing this will allow every one of us to live a full life, mentally alert and physically active, well into our nineties. When that happens, it will be the biggest paradigm shift in the history of medicine—from a disease to a wellness focus.

There are hurdles ahead of us. Lots of them. Some are scientific, but far more are psychological, sociological, and economic. Who will have access to this data-intensive next-generation medicine, and who will be left out? Will we be able to make it easier to escape the poor health choices that provide immediate pleasures but bear steep negative penalties in the long run? These are big questions that are yet to be answered, but at the moment the single greatest barrier to transforming this vision into reality is the need to persuade healthcare systems (providers) and payers to accept a dramatic change from a model built on disease to a paradigm based on wellness. But that, too, is an achievable goal. If it were not, this book would be little more than a gee-whiz appreciation for new science accompanied by hazy speculation about where it might lead us in the future. We don't see it that way. We see it as a clear plan. Indeed, we see it as a plan for a movement.

We are fortunate to be at the forefront of that movement, as both of us are leading independent efforts while continuing to collaborate on research at ISB and integrating our findings into a shared view of what the future of health should be.

Lee's primary focus since 2021 is on a project he is leading called the Human Phenome Initiative. This is an ambitious plan conceived by Phenome Health, a nonprofit organization Lee created to advance the science of wellness and prevention and pioneer a data-driven approach to optimizing the mental and physical health of everyone who wishes to participate.

The aim of the project is to start with one million individuals selected to reflect the racial and economic diversity of the United States. It is an enormously ambitious endeavor, one that we believe will demonstrate to the world the power of P4 healthcare, or what we call N = 1 medicine—the idea that, while vast troves of data can and should be used to inform possibilities for extending wellness and reversing disease, no doctor should ever gamble with a patient's life on the shopworn notion that a treatment that is good enough for some people, some of the time, is right for anyone who shows up, without taking account of their specific genetic and biological makeup. As a global medical community, we can do better. It's long past time for us to do so.

For Nathan, being at the forefront of a global embrace of scientific wellness has meant taking on the role of chief scientific officer for Thorne HealthTech, a public company focused on personalized scientific wellness that is using individualized testing, AI, data, and efficacy trials to create products and services to make healthy aging accessible. Thorne is currently working with five million consumers and more than 47,000 healthcare providers. Nathan is particularly focused on implementing brain health solutions (building on the science described in Chapter 8) and creating multiomic wellness tests—including those derived from metabolomes, proteomes, microbiomes, and immune-cell profiling—to provide insights into the workings of our various biological systems.

We believe that our efforts—and those of many other physicians, scientists, naturopathic doctors, philanthropists, entrepreneurs, and public health advocates who believe as we do that the healthcare system needs to be reinvented to focus first and foremost on wellness—will help validate the central hypothesis of this book: that a revolutionary restructuring of healthcare from a model that is centered on disease to one that is centered on wellness is both necessary and overdue.

So how do we get there?

Making Wellness Last—The Central Concept

The health trajectory of each individual has three distinct phases: wellness, transition, and disease. The data-driven approach to scientific wellness will optimize health far beyond what most people think is possible, provide us with approaches to track and slow aging, identify transitions to disease years before clinical diagnosis, and apply systems-driven approaches to slow or reverse nascent disease at that early and simple stage. It will also inform pioneering data-driven observational studies to delineate the different subtypes of most chronic diseases, understand the biomarkers that identify these transitions, and find targets for therapy to block disease at these different transition states. Precision medicine generally applies to the search for biomarkers and therapies only after clinical disease diagnosis. Hence most precision medicine, as currently practiced, happens far too late, because it occurs once a disease has progressed to irreversibility.

There is no known biological law that dictates when transitions to disease must arrive in our life spans. For most people in the United States, United Kingdom, Western Europe, and many Asian countries, where sedentary habits, high stress, and poor diet are taking a particular toll, the transition to disease generally begins in one's thirties or forties and accelerates in the fifties and sixties. Above age 55, 78 percent of Americans have been diagnosed with a chronic disease, nearly half have been diagnosed with at least two, and one-fifth have three or more diagnosed chronic diseases.[1] For Americans over 65, as many as 86 percent have been diagnosed with one or more chronic diseases, more than half with two or more, and nearly a quarter with three or more. But together with a data-driven approach to health, we can create a world in which this transition to disease no longer plays out as it commonly does today.

This idea is highlighted in a figure you may recall from the opening chapter of this book, which we offer here again because it is paramount to understanding the path we must take to implement scientific wellness at a global scale (see Figure 11.1).

The key change from the healthcare of today to the healthcare of the future hinges on the optimization of individual wellness and early detection of wellness-to-disease transitions, offering the potential for reversal before the emergence of clinically diagnosable disease. The progression of disease leads to ever greater disease complexity as growing numbers of biological systems become perturbed. If we wait—as we now do—for symptoms to appear, reversing the course of disease becomes infinitely more complicated. An approach focused on early detection will permit a much longer phase of inherent wellness, punctuated

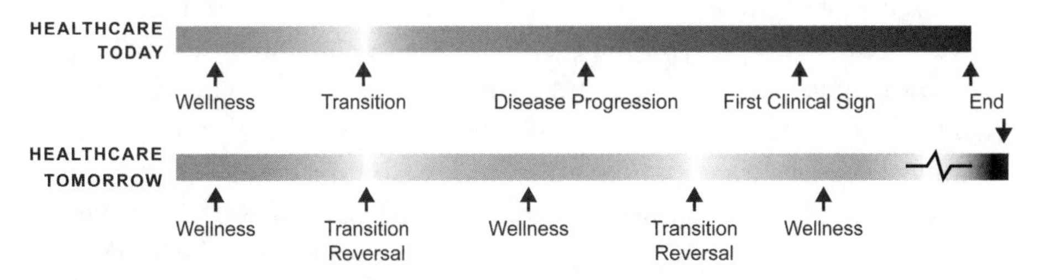

Figure 11.1. Treating early transitions will greatly extend our health span, bringing down costs and improving lives.

by brief and perhaps even imperceptible periods of transition and reversal.

This strategy will necessarily include regular measurements, starting with whole-genome sequencing and continuing with the tracking of multiomic health data including blood analytes, the gut microbiome, and digital measures of heath, all of which will become components of each individual's longitudinal data cloud. With these data, we can understand the full spectrum of risk for thousands of diseases and assess strategies for reducing these risks across a lifetime, enabling healthy aging in both body and brain. Wellness won't be something we only think about at the time of yearly checkups with our doctors but, rather, a fundamental part of our day-to-day lives. We will make choices and develop habits around how we sleep, eat, exercise, and assess our blood, gut, and digital data to meet our immediate health needs and optimize our potential for a future rich in wellness.

While caring for our bodies, we will not ignore our minds. Exercise and better sleep management, coupled with cognitive assessments and training to make the most of the intrinsic neural plasticity of the brain, will help us prevent or reverse any loss of cognitive function.

It is obvious that the longer we can make the wellness phase last, the better. But how long can this period go on? We cannot say for certain, for there has not yet been a human being in their eighties or nineties who has enjoyed a full life of scientific wellness, and there won't be for a long time. A child today who has these systems and supports in place before entering the first years of healthy adulthood might just find themselves enjoying a state of physical health and mental agility in their eighties that is virtually imperceptible from the decades before.

If we can live functionally into our nineties or 100s, what will the experience of death be like? Right now, for most people, it comes after a years-long period of decline, increasing disease, and debilitation. Yet most people who do make it into their 100s do so in relatively good health and generally die quickly of whole-systems failure—thus promising a mercifully quick, relatively painless, fortune-conserving end of life. Most people don't like to think about death at all, but this is the kind of end we should all hope for—and the kind that a healthcare system focused on wellness would help us achieve.

We must acknowledge that there will still be disease. Over the course of a lifetime, the human body and mind will move through a minefield of genetic predispositions, injuries, stressors, toxins, radiation, poor lifestyle choices, traumas, and pathogens, all of which can contribute to disease progression, even for those who have done everything they can to live in a state of wellness—and even more for those who have not. But unlike the way we experience disease today, the deep monitoring of scientific wellness will give us meaningful signals to ensure that in the future most transitions can be turned around in their earliest states, and those that cannot will be detectable and managed many years before symptoms manifest themselves.

When disease does manifest, we will confront it with the powerful tools of precision medicine brought to us by systems approaches, big-data observational studies, machine learning, and artificial intelligence. The power to treat patients effectively will continually increase, as precision medicine will power systems analyses of both well-known and new diseases, providing insights into diagnosis, the stratification of diseases into subtypes, the delineation of the transitions states for each subtype, and the development of new multimodal therapies specific for each subtype and their transitions. What this future means is that, when we are diagnosed with a serious disease, it is possible and even likely that we will return to wellness.

So how do we bring this vision of scientific wellness and prevention to reality?

The Collision of Systems Biology and Scientific Wellness with a Large Healthcare System

Over the course of his career, Lee has provided tools—starting with DNA sequencing, protein sequencing, and DNA synthesis—that make it possible to generate big data on individuals. The Human Genome Project gave us the ability to correlate genetic variability with wellness and disease phenotypes, though it took us much longer than anticipated to effectively begin to tease out these correlations. Systems biology offered a new holistic, dynamic, and hierarchical view of the networks governing biology and disease, catalyzing the vision of P4 medicine and triggering the use of computational tools such as knowledge graphs, machine learning, AI, and digital twins. The power of data-driven population health was made clear by the Pioneer 100 and Arivale initiatives, leading us to the science of wellness and prevention with its power to optimize individual health trajectories. But how can these ideas take hold given the reality of our modern healthcare systems that are focused almost entirely on disease?

In 2016, the CEO of the Providence healthcare system, Rod Hochman, approached Lee with an irresistible offer: the position as the chief science officer for Providence, a role that would offer the opportunity to bring scientific wellness and systems biology into our healthcare system. At the time, Providence had fifty-one hospitals in seven western states, treating more than five million patients. This was an ideal opportunity to test the challenge of bringing scientific wellness to doctors and patients in the context of an existing hospital system. Lee jumped at the opportunity. A few months later, he piloted a program, meeting every three months, involving twenty-seven clinical and research physicians from across the Providence system who were introduced to research that centered on the power of systems biology and scientific wellness. He focused on the idea of bringing dense-data analysis of the genome and longitudinal phenome into clinical trials, which sounded great in principle but had never actually been done in a clinical context before.

Most receptive to this idea was a group at Providence in Portland, headed up by Dr. Walter Urba, an excellent oncologist specializing in melanoma who was advancing precision medicine and the use of DNA sequencing to develop more individualized drug treatments for cancer. In addition, during the last two years of Arivale, human geneticist Dr. Ora Gordon headed up a scientific wellness program with 1,000 Providence employees, which unfortunately terminated early, after two years, with the premature termination of Arivale. Just as it took time to build support and enthusiasm for DNA sequencing, it will take time to convince a critical mass of physicians and researchers to embrace an ongoing program of scientific wellness, but with passionate advocacy from a few respected physicians like Urba and Gordon, it's clear we are on our way.

COVID-19 Propels Innovation in Precision Medicine

The other great clinical breakthrough came, unexpectedly, from research on COVID-19.

There was a great deal we didn't know about COVID-19 in the spring of 2020, when novel and deadly variants of the coronavirus began to circulate across the United States, but one thing was immediately clear. The disease did not strike everyone in the same way. Older patients were more susceptible than younger ones, men more than women, and some racial groups and blood types were more vulnerable than others. Little was understood about how the immune system responded to the virus, and there was widespread uncertainty as to which drugs would be effective to treat it.

At ISB, Jim Heath was determined to play a role in answering these questions. Dr. Jason Goldman of the Swedish Medical Center, the largest Providence Health–affiliated hospital system in Seattle, worked with Jim and other colleagues at ISB to apply genome and longitudinal phenome analyses for the purpose of better understanding the disparate outcomes and clinical pathways of COVID-19 patients. The team coupled these analyses with an important new tool—deep immune phenotyping—which made it possible to describe the state of a patient's immune and biological responses at any given point during the infection. This led to an observational study, starting in March 2020, including about 200 patients, which followed the course of this disease in individuals as it progressed after admission to the hospital. What the team ultimately discovered offered substantial new insights into COVID-19-triggered immune responses. Perhaps as important, the study catalyzed a series of new opportunities to advance medical practices across many fields of study and care. These reflected a broad, if begrudging, acceptance of outside-the-box thinking during the COVID-19 crisis that has begun to permeate global medicine, despite the fact that it runs contrary to the "this is the only way of doing things" mindset so pervasive in the contemporary medical research system.

The ISB team was able to set up this data-rich COVID-19 study in a remarkably short time—circumventing the typical bureaucracy of large healthcare systems to accomplish in weeks what would normally have required years of planning. Finding funding for the project—a sizable $10 million—proved surprisingly simple. Roger Perlmutter, an insightful immunologist who was president of Merck Research Laboratories, needed little more than half an hour to decide that deep immunological phenotyping would be a good bet for his company and, more important, for the world. He pushed for extreme transparency so that other scientists could help decipher the data and contribute to unraveling the mysteries of COVID-19 without the sorts of intellectual property constraints that generally prevent this type of cooperation. This was unusual for a project funded by a pharmaceutical company. Several other pharma companies, including Novartis and Gilead, soon joined the study under similar conditions, and about ten smaller companies did as well, contributing novel technologies for immune data generation. All this collaboration—and the esprit de corps that ran across the project—created an atmosphere in which it was easy to recruit academic experts in infectious disease and immunology from many outstanding institutions across the United States to join us in interpreting this valuable and dense data set. This is the way science should be practiced.

The strategy was to obtain blood from each patient at admission, a second sample about ten days later, and a third up to three months later. These three blood draws gave us a chance to study the nature of the infection at admission, to see its evolution to a point after the acute infection had run its course, and to benchmark where the immune system stood at convalescence—the period of time after the acute disease when someone is recovering health and strength, gradually, following their illness.

We conducted complete genome sequencing and longitudinal phenome analyses of about 500 blood proteins and 1,000 blood metabolites. We also carried out deep immune phenotyping, with single-cell analyses on 5,000 single white blood cells, at each blood draw—analyzing the complete transcriptome (the expression of all genes), measuring 250 cell-surface proteins, and investigating forty secreted molecules from each cell. We then analyzed each patient's T cells and B cells receptors, identified the antigen-presenting HLA locus, sequenced the viral genomes at various stages of the disease, and made a library of viral peptide fragments, called epitopes, to determine which ones the antigen-presenting cells used to trigger killer T cells. The deep immune phenotyping allowed us to identify each white blood cell type and its stage of differentiation. When the data from the 5,000 immune cells of each single blood draw were integrated, they defined the state of the adaptive and innate immune responses at each of these three time points.

This analysis gave us a detailed view of the path of the viral infection, helping us understand patients' health trajectories and the range of divergent responses to various drugs. We emerged with profound insights into the nature of the immune response throughout the infection. And we were able to define new classes of T and B cells at various stages of the disease and to conduct single-cell analyses, assessing the metabolism of these cells and once again revealing interesting new classes of immune cells. Our observations were published in three papers in prominent journals—two in *Cell* and one in *Nature Biotechnology*.[2]

Comparing immune cells at the time of admission and ten days later allowed us to capture striking changes in individual immune cell types at different time points and helped us see how different immune systems responded dynamically to drugs. This helped us think about how to design optimal clinical trials for patients undergoing treatment with different drugs.

The data at admission allowed us to divide patients into four groups (two with mild disease and two with severe disease) by both immune profiles and blood protein biomarkers. The patients in only one of these groups saw their immune systems return to normal three months after admission, the normals representing only about one-third of the total patients—suggesting that the immune responses in the others played a key role in the prolonged "long COVID" disease, possibly affecting the nervous system with brain fog, loss of smell, and fatigue, and leading in some cases to abnormalities in respiratory, cardiac, or gastrointestinal physiology. We found that so-called "long COVID" was in some instances relatively short in duration (weeks to months); in others, it lasted considerably longer.

In an early stage of the disease, we found in the blood work a striking loss of essential amino acids and phospholipids, both components of the building blocks for immune cells. It may be that the rapid expansion of immune cells leads to this temporary deficiency. We also found that many of the symptoms of long COVID were autoimmune in nature—where the immune system reacted against its own components.

How the different types of long COVID arise is still a mystery, but we now know much more about the factors that play into these disparate experiences. For example, we identified four key factors at hospital admission, any one of which suggested that a patient would get long COVID.

• The first factor was pre-existing type 2 diabetes. That's fairly common knowledge now, but this insight played a critical role in helping us identify those who were at greatest risk.
• A second factor was the presence of certain autoantibodies—blood proteins that attack the body, which are common in people with all autoimmune diseases including those like lupus or rheumatoid arthritis as well as less and more common autoimmune conditions.
• A third factor was the magnitude of COVID-19 RNA located in a patient's blood right at admission. Everyone had it in their saliva. But we found that the roughly 25 percent of patients who had it in their blood had more severe outcomes.
• The last factor was the activation of another latent virus, the Epstein-Barr virus, a form of herpes virus that is relatively common and best known for triggering mononucleosis. This virus appeared to have been "reawakened" by COVID-19 in about 14 percent of the patients, who were then far more likely to develop long COVID symptoms.[3]

All of these findings proved valuable in helping identify patients who were more likely to need aggressive treatment, more watchful monitoring, and an earlier place in line when the first FDA-approved vaccines became available.

In the long run, our data-rich COVID-19 observational study showed that, with genome and phenome analysis and deep immune phenotyping, scientists can draw powerful conclusions from a limited number of patients. Well-designed studies of a few hundred people may no longer need to be considered "small," "incomplete," and "inconclusive." We showed that we can identify a disease we want to understand, investigate its trajectory in a few hundred patients with the tools of whole genome sequencing and deep phenotyping, and quickly emerge with a wealth of knowledge about its trajectory, giving us a chance to exploit avenues for early intervention that would have been unimaginable just a few years ago.

More broadly, the COVID-19 pandemic revealed how vitally important a current healthy state is in promoting resilience to disease. Pre-existing conditions greatly predicted who would suffer the most and who was most likely to die. A 2021 meta-analysis across 160 primary studies considering forty-two disease conditions revealed that, compared to respective control groups, diabetes increased the risk of dying from COVID-19 by between 20 to 100 percent, obesity by 50 to 75 percent, heart failure by 30 to 130 percent, chronic obstructive pulmonary disease by 12 to 120 percent, and dementia from 40 to 670 percent across different studies in different regions of the world.[4] This study also reported that people in Europe and North America further showed increased risk of dying from COVID-19, with liver cirrhosis increasing the risk by 220 to 490 percent and an active cancer by 60 to 370 percent. Clearly, pre-existing health was the dominant factor in whether someone was likely to die from COVID-19, and you may have noticed that this is essentially a catalog of the most common chronic diseases in the world. These diseases are also vastly ameliorable through scientific wellness.

A study published in January 2022 highlighted not only risk factors but also *protective* factors associated with less disease severity and death from COVID-19. These protective factors obviously included vaccination, but healthy diets with sufficient nutrition were also powerfully protective.[5] For example, vitamin D consumption was particularly associated with reduced viral replication rates and reduced levels of pro-inflammatory cytokines, the immune-signaling molecules that circulate throughout the bloodstream, as well as reduced risk of infection and death.[6] Adequate vitamin D may also help protect the lining of the lung from pathogenic invasion.[7] Natural products such as quercetin, abundant in food sources including dark berries, were also associated with reduced risk and improved therapy outcomes in clinical trials.[8] From these and other findings, it's clear that the reduction of disease and accentuation of healthfulness—both inherent parts of scientific wellness—can play a combinatorial role in assuring better outcomes in a pandemic such as this.

Research has also revealed strikingly critical differences in how people of different ethnicities fared in the pandemic.[9] The literature on this subject is complex and still evolving, but a meta-analysis of fifty quality-screened primary studies published in November 2021 found a number of instructive differences.[10] Black and Hispanic patients had the highest risk of being diagnosed with COVID-19 and higher disease severity on average; however, this analysis did not show a higher risk of dying. Asian Americans were the most likely to end up in an intensive care unit. During the time when omicron was the dominant variant, peak hospitalization rates among Black Americans were nearly four times the rate of white Americans, and the variants dominant at any given time impacted outcomes.[11]

There are many reasons for these discrepancies, and a continuing flurry of research seeks to makes sense of them. Socioeconomic differences played a key role, accounting as they do for differential access to high-quality healthcare and a higher likelihood of people performing in-person jobs at the height of the pandemic. Another big reason for the discrepancy is differences in vaccine hesitancy. A 2021 study in Arkansas found that Black respondents were 2.4 times more likely to be resistant to pursuing vaccination than white respondents.[12] A meta-analysis of forty-five primary studies across the United States and United Kingdom, which evaluated ethnic discrepancies in the context of comorbidities, found that unadjusted all-cause mortality during 2020 was about the same for white, Black, and Asian ethnicities, with Hispanics having a lower mortality rate.[13] However, after adjusting for age and sex (white people are significantly older on average than minority populations in the United States), Black patients had a 38 percent higher chance of dying from COVID, and Asian patients a 42 percent higher chance of dying than white and Hispanic patients. After adjusting for existing health conditions, researchers found in this study that the differences in outcomes during the first year of the pandemic, prior to vaccines, seemed to be mostly explained by pre-existing health conditions. Individual health was shown to be absolutely critical to the likelihood of resilience. This was, once again, a validation of the central thesis of scientific wellness and a shift from disease management to the preservation of wellness.

We would, of course, trade all this knowledge for the millions of lives lost to this disease around the globe. But as we cannot rewrite the past, if the epilogue of this pandemic is that it helped us see the folly of old ways of doing healthcare and allowed us to better recognize the opportunities inherent in a new path, then at least it can be said that we are learning. And, indeed, the striking success of our COVID-19 observational study almost immediately sparked similar efforts at Providence to better understand multiple sclerosis and breast cancer, incorporating both systems biology and the power of multiomic data analyses.

Genomics: Leading the Way

Of all the -omics technologies, genomics is the first that is being integrated at scale into healthcare systems. The widest adoption of genomics today is for sequencing tumors and comparing these sequences to the genome a patient was born with. This information is then used to target therapies with a higher likelihood of success, as we discussed in Chapter 9. Genomics as a basic piece of everyone's healthcare is coming, but it hasn't come as quickly as those involved in the Human Genome Project originally hoped, and that has fueled critics, who have argued that genetic sequencing and analysis is not actionable, that it identifies extremely rare conditions at great cost, that knowing one has a genetic predisposition for some terrible disease will cause anxiety, that participants or their healthcare providers will misunderstand genomic information and inadvertently cause harm, that genomic-based medicine is not cost-effective, and that privacy concerns cannot be satisfactorily resolved.

These are not concerns to be blithely ignored. But in most cases, they represent a fundamental misunderstanding of the data, the trajectory of innovation, and the needs of patients. Pioneering genomics researcher Robert Green, a professor at Harvard Medical School and director of the Genomes to People project established in partnership with the Broad Institute, set out to perform a series of studies to investigate each of these issues in detail, some of them funded by the National Institutes of Health. And while privacy concerns remain important and have not been adequately addressed in many respects, for the most part, none of the other fears has been supported in rigorous studies. Let's discuss these in turn, since this work is critically important as we undertake a fundamental—and irreversible—step toward using our DNA to inform healthcare.

When the first genome was sequenced, it was true that genetic codes didn't provide much actionable information. A genome at that time was like a reference library that is full of books but devoid of any "how to" guides to put knowledge into context and offer a plan of action. As the number of genomes has accumulated, and the sharing and comparisons of these codes has become possible, this has slowly begun to change. As of 2022, there were at least seventy-six genetic variants classified as clinically actionable based on standards set by the American College of Medical Genetics, including for Lynch syndrome (colon cancer), breast cancer, ovarian cancer, heart failure, and sudden cardiac death.[14] As of 2019, there were also at least 132 pharmacogenomic variants that can be used to inform dosing or usage of ninety-nine drugs; this information is now included in 309 medication labels.[15] These pharmacogenomics variants include those used to guide treatment for breast and colon cancer, such as whether a patient can be safely prescribed the chemotherapy drug 5-fluorouracil. Another example is human leukocyte antigen testing, which is now required for prescribing Abacavir, one of the key drugs used to treat HIV. Pharmacogenomic information can be absolutely vital, as response to drugs like these in individuals with incompatible genetics can be lethal.

One limitation of single-genetic variants for common diseases is that their effect size is often small, with the vast majority having less than a 1 percent effect. The evolutionarily commonsensical but brutal reason for this is that these variants wouldn't be common if they were highly detrimental to survival. Thus, for common chronic diseases, genetic variants are combined into polygenic risk scores, an aggregate score with far greater predictive power than any individual variant. With your genome sequenced, these scores sum up the aggregate risk you have for a disease based on the contributions of many individual variants that have each been associated with a particular disease or condition. There are more than one hundred conditions for which well-validated polygenic risk scores have been developed, and this number is growing rapidly. This information can provide doctors and individuals with information about a wide spectrum of potential predispositions, ranging from stroke and anxiety to Alzheimer's and diabetes.[16] If a person knows their whole-genome sequence, polygenic scores can be converted into disease risk scores. The combination of genomics with other data types—including clinical, wearable metrics, and multiomic deep phenotyping data—considerably increases actionability.

One important caveat for the use of polygenic risk scores is that they are heavily biased by the makeup of the populations that have been most frequently examined in genetic studies to date, and are thus most accurate and applicable for white individuals.[17] This is absolutely true, and although racial heterogeneity of data is improving, it will still be some years before enough genomes are sequenced to provide the same degree of accuracy across most ethnicities.

A second objection is that, while genomics are clearly useful for investigating rare diseases—impacting fewer than one individual in 200,000—those diseases, by definition, are not common enough to be a major factor in the overall health of a population. Rare diseases are often caused by single-gene defects rather than by collective effects across common variants of environmental triggers. These are much easier to find and thus lend themselves to being discovered through whole-genome sequencing.

The Hood Lab carried out the first genome sequencing of a family with a genetic disease in 2008. By comparing the genomes of the parents with those of their two children (one of whom was affected, the other not), we were able to identify the genetic abnormality responsible for Miller's syndrome, a finding published in *Science* in 2010.[18] Since that time, numerous family genomic sequencing efforts have been undertaken, using genetics to track down rare disease genes and pair them with repurposed drugs to improve treatment outcomes.[19] What we have learned over the past decade is that, while each rare disease is indeed rare, in aggregate, they are not so rare; 10 percent of us will suffer from one or more of the 7,000 known "rare" diseases at some point in our lifetime.[20] Also, because uncommon genetic diseases often manifest themselves early in development, there are major efforts under way to identify and treat these conditions in infancy or early childhood, informed by genome sequencing, as 20 percent of infant deaths are tied to genetic abnormalities.[21]

A third major concern is that knowing one has a genetic predisposition for a given disease would cause undue anxiety. While this is a valid idea to study, it has long been taken as self-evident—an unscientific approach to any idea, much less one being used to prevent people from accessing information about their own genetic selves. Upon examination, the hypothesis simply hasn't held up. In a study of people using commercial genetics testing published in *Public Health Genomics* in 2017, researchers found that only 2 percent reported regret after receiving their results, and only 1 percent reported that the results had caused them any actual harm.[22] Another randomized clinical trial evaluated the effects of informing patients about APOE, the gene linked to differential Alzheimer's disease risk and one of the disease-predisposing genes brought up most frequently by critics who claim that more knowledge will bring needless anguish to patients. The study's authors found, however, that providing this information to patients rarely generated anxiety.[23]

Since broad genomic screening generally finds disease risks that are both elevated and lower for the vast majority of people, overall, the bad news is generally offset by good news. But even for those for whom the news is abjectly bad, it's also increasingly coupled with actionability—which is to say that there are specific steps you can take to avoid potential disease in the future. If you learn that you have high genetic risk for type 2 diabetes, for instance, you can be particularly vigilant about controlling sugar intake, consume a high degree of dietary fiber to reduce glucose spikes, begin supplementing your diet with natural compounds such as berberine, fortify your gut flora with the probiotic *Akkermansia*, or begin taking a medication such as metformin in consultation with your physician. If you have a high genetic risk for Alzheimer's, you may help delay dementia through exercises that keep your brain oxygen perfusion high, and you may consider taking a supplement such as phosphatidylcholine and engaging in regular digital cognitive exercises such as BrainHQ. Anxiety is rarely occasioned by bad circumstances alone—it is far more often invoked by circumstances that you can't do anything about. These days, though, there's a lot that we know we can do to mitigate disease risk, and we're learning more every single day.

The fourth objection is that patients or their providers, most of whom know very little about genomics, will simply not be able to properly interpret the genomic information. While all these concerns are valid starting places for study, this is the one we disagree with most of all. To be frank, it's alarmingly condescending. We believe people should make their own informed choices about what kind of information they want or don't want. We should nonetheless hold genetic companies and genome-informed healthcare providers accountable for explaining the complexities of genomics analyses for making their insights and recommendations as clear and accurate as possible. Genetic counselors and genomics-educated physicians have an important role to play in helping people to understand this information deeply.[24]

The costs associated with the effective deployment of genomics in healthcare systems is another often-stated objection—but even the skeptics in this case are mostly arguing in terms of time, since the costs associated with sequencing the genome have fallen so dramatically—a millionfold over the past two decades. Today, the cost of whole-genome sequencing in many places has fallen to less than $400, and while that might seem like a lot of money when spread across a population, in reality, it doesn't cost much more than many of the standard blood tests that doctors regularly order for their patients—and in some cases, it's a whole lot less. And aside from the vast wealth of information that we can glean from each person's genome sequence, there's one other major difference: it only has to be done once.

Now we arrive at the matter of privacy. And here, as we explained in Chapter 4, there are definitely legitimate issues. It is important that safeguards be put into place to regulate the use of personal genetic information. The biggest difficulty with genomic privacy is that your genetic code is unique to you. In that sense, it can never be completely anonymized. There are strict laws for maintaining the privacy of genetics data and the allowed uses of such data, as codified by HIPAA in the United States, but this has not prevented flagrant violations and the creeping use of genetic information that would make many people quite uncomfortable. It's important to note, however, that at this point more than 20 million people have carried out genomic analyses commercially—and nefarious uses of genetic information remain exceptionally rare. It's nonetheless clear that we will always be fighting a battle to protect privacy and that this is an issue that will require vigilance and care. It is crucial that appropriate legal safeguards be put in place and that we never assume that rules written in the past will safeguard the future.

With evidence accumulating to debunk most of the common concerns about the integration of genomics into mainline clinical practice, a number of the nation's top healthcare systems are finally moving to make this long-awaited aspect of care a reality. Within the next five years, the genome will become a basic tool in leading healthcare systems. Within ten years, it will be an almost ubiquitous part of the healthcare environment. It will be used not just for advanced treatment of rare diseases or for cancers, as it is now, but for prevention. One study of note in this regard is the BabySeq project, led by Drs. Robert Green and Ingrid Holm of

Harvard Medical School and Boston Children's Hospital.[25] BabySeq, part of the Genomes to People Project, is demonstrating the value of genome sequencing starting from the very beginning of life, exploring critical issues such as improving health outcomes for babies, parental attitudes regarding standard newborn genomic sequencing, and the perception of benefits and risks.[26] Hospital-sponsored studies like this are critical as we move into a world of scientific wellness, with its goal of delivering genetically and phenomically informed preventative healthcare.

If this sort of information excites you right now, it is probably of little solace to know that it is coming at some point down the line. But if you are determined to access this kind of care now and have the resources to supplement traditional care, you don't have to wait that long.

Providence and the Genome4Me Program

As a part of Lee's efforts to bring genomics and scientific wellness to Providence, he lobbied for a combined genome and phenome approach to scientific wellness, similar to what we implemented at Arivale.

The then chief clinical officer at Providence, Dr. Amy Compton-Phillips, expressed considerable enthusiasm for genome analysis, and under her leadership, the system has initiated the Genome4Me program, using whole-genome sequencing on cancer and control patients under the direction of Walter Urba and his colleagues in Portland, Oregon. Providence set a goal of recruiting 1,000 patients in the first year and 5,000 over the next several years; the Portland group acquired the necessary sequencing equipment, and the program kicked off in 2021. The idea was that patients and their doctors would be presented with actionable recommendations based on information in their genomes. This effort took seriously the challenge of educating both patients and physicians. One attractive feature was that much of the cost of genomic analyses would be covered by insurance, as the patients had already been diagnosed with cancer and sequencing would inform their treatment possibilities. And it placed Providence in the front rank of US healthcare systems practicing genomic medicine.

Even at Providence, where leadership is unquestionably forward-thinking, implementing widescale phenomic analyses has been a challenge because no insurance companies, at this point, are willing to pay the costs, save for when it comes to blood chemistries that already show known disease relevance.

There have been other concerns, all of which will be important to address. Even if tests are paid for, who would be responsible for paying for the time of physicians educating patients about these possibilities? How can doctors be trained to understand this complex new genomic science when they are already stretched so thin taking care of their patients?

The path to scientific revolution—to the Big Sky moments we described in the introduction to this book—is to think very big but be willing to celebrate any meaningful steps toward your goal, no matter how small those steps seem compared to the extent of your desires. To this end, while we would obviously love to see a full-blown genomic and longitudinal phenome program, we are enthusiastic that Providence has been working to de-identify patient data and, having taken this step, to protect patient privacy. It has also built partnerships with pharmacological researchers so that they can investigate the clinical features and stages of various diseases en masse. Efforts like these will benefit from another Providence program: as part of its Genome4Me program, Providence is now providing patients information on their vulnerability to more than seventy variants that have been identified as actionable by the American College of Medical Genetics and Genomics, several pharmacogenomic variants that are known to interfere with the effective use of drugs, and precision cancer drugs.

To date, cancers are the area where genomics has been most widely employed, which simply makes good sense. One notable example is the blood tests known as liquid biopsies that make it possible to identify circulating cancer cells or pieces of DNA from those cells. It's increasingly possible to discover cancers in the treatable early stages, rather than the more advanced Stage 3 or metastatic Stage 4.[27] The evidence that tests like this can have a big impact on care is mounting. One study showed that patients lived significantly longer and treatment cost dramatically less if a precision medicine approach was used for their treatment, saving hundreds of dollars a week per patient.[28] In the Providence system and other networks across the globe, these savings are starting to add up, and the inevitable result will be greater access to a fundamental element of scientific wellness—genomic actionability.

Other Examples of Genomics in the Clinic: The Beginnings of a Transformation

While Providence has been a leader in these areas, they haven't been alone. Genomics is starting to be incorporated across leading healthcare centers and is poised—at long last—to become a routine part of healthcare. Another healthcare system that has started this implementation in earnest is Intermountain Health, the largest healthcare system in Utah, where Lincoln Nadauld, chief of precision medicine, has spearheaded a large-scale genomics effort that has sequenced more than 125,000 patients. About 7 percent of these patients have been identified to have an actionable variant, including genes indicating elevated risk for breast cancer, ovarian cancer, cardiovascular disease, hemochromatosis, and seizures. These patients—and their doctors—now have information that they can use to make decisions about lifestyles and treatments, and the end result won't be just better health but tremendous savings. As Nadauld put it, "Many clinical screening tests that we regularly perform in healthcare today have less than a 7 percent positive hit rate. This study showed we should actually just sequence everyone—and health plans should be interested in paying for it because they are the ones that financially benefit."

Intermountain Health has also implemented whole-genome sequencing for infants in neonatal intensive care units, and it turns out that half of these babies had a genetic defect that explained why they weren't improving—information that was in many cases actionable for care providers. This knowledge markedly reduced intensive care, resulting in significant cost savings for the healthcare system and, quite certainly, better lifetime health outcomes for the newborns, as a child's early days are so critical to their wellness for the rest of their lives.

With such promising results, it is a wonder that every healthcare system in the world isn't rushing to implement these elements of scientific wellness. Alas, as has often been observed, big ships turn slowly—and healthcare is a very, very big ship.

Large-Scale Discovery Efforts

Most of the advanced healthcare systems in the world are single-payer systems funded by governments. In contrast, the United States has a maze of different payers spread out across the country. If one wishes to alter the healthcare system, it is far easier to do so under a single-payer system than one like the US system, where each healthcare provider and each payer has to be independently convinced to change.

There are now several ambitious government-funded research projects underway that are generating large-scale data sets to help advance discoveries that will aid in the rollout of personalized and predictive medicine. These projects are primarily focused on genomics, but their banked blood samples could also be used down the road for deep phenotyping.

A leading example of this is the 500,000-person United Kingdom Biobank, which has generated most of the recent large polygenic risk scores for disease associations.[29] A proteomics-focused extension of this program recently captured and characterized the blood proteomics profiles of more than 54,000 people in collaboration with thirteen biopharmaceutical companies. This first-phase measured nearly 1,500 blood proteins in each sample and identified more than 10,000 genetic associations with the observed levels of these proteins in the blood. Fully 85 percent of these associations were previously unknown. This effort has given medical researchers a view into the relationships between genetics and protein levels at an unprecedented scale.[30]

Results such as these are compelling evidence of the need for wider-scale projects, like the UK's Our Future Health program, which intends to carry out genomic analyses on five million people and integrate that fundamental information with patient digital health measurements and electronic health records. As additional testing data is added, especially as other omics data types such as metabolomics are added longitudinally, the scale and scope of discovery will transform our understanding of human biology and medicine.

The United Kingdom is not the only single-payer nation moving in this direction. In Singapore, a project that aims to sequence the genome of one million Singaporeans was launched in 2020; it is expected to take ten years to complete this effort, and before the end it will include some longitudinal phenomic analyses. Even the far smaller country of Estonia has a million-person project underway, with an attendant biobank for samples initially focused on genomics, EHRs, and digital health measurements that will essentially cover every resident of that nation.

In China, perhaps the country that is most advanced in multiomic phenomics, a large-scale population program is also underway. Managed by Professor Jin Li, the president of Fudan University, the project launched with an initial registry seeking 10,000 individuals and its leaders hope to quickly expand it to half a million or more. To make that possible, they have established their own data-generation facilities for genome, multiomic, and other multidimensional phenome analyses. In addition, they integrated into their program electronic health records and patient reported outcomes. They also have incredibly complete multidimensional imaging facilities both for body and brain. Moreover, they are assessing techniques for measuring many different aspects of the environment.

The Fudan project is being closely watched by the Chinese government, which initially committed more than $100 million US dollars to the genome and phenome mapping effort and the infrastructure needed to support it. Other nations should take note: this data-driven, precision population health program is included in one of the five major scientific efforts that Chinese leaders have initiated—and supported with striking resources for rapid development—with the intent of achieving global leadership by 2030.

As single-payer nations move quickly to embrace this aspect of scientific wellness, will the United States be left behind? That's possible, but all is not lost. The National Institutes of Health's All of Us program, first proposed by a working group advising the institute's director of precision medicine, is a research effort aimed at generating genomic data for one million people and linking this with medical and health information, especially electronic medical records, in an effort to pave the way to the wider dissemination of personalized medicine.[31] All of Us hopes to scale precision medicine and to help move the US healthcare system beyond its current disease-specific, "trial and error" approach, which treats every individual as if they were an "average patient." One key focus is the diversification of the patient pool to better inform thousands of studies on a variety of health conditions, understand the risk factors for many diseases, and determine which treatments work best for people of different backgrounds. This highlights the participatory nature of the future of medicine in its stated goal of learning "how technologies can help us take steps to be healthier." While the project aligns with our vision of data-rich, genomically informed healthcare, it is so far missing the element of longitudinal proteomics and the return of actionable possibilities to individuals. However, because All of Us is banking samples and making these available to researchers for further studies backed by subsequent NIH grants, phenomic data will grow more abundant with time, especially if the sample collections become longitudinal.

Seven Challenges

Our experiences at Arivale, partnership with Providence, and work with the research community seeking to improve population-wide health outcomes has led us to conclude that there are seven major challenges to validating and establishing the science of wellness and prevention.

How Can We Include Everyone in This Vision of Wellness?

Foremost, we must recruit individuals from every demographic group in the United States. That will necessarily include gender, age, and racial diversity, of course, but also economic, cultural, geographic, educational, occupational, religious, sexual, familial, and ability diversity. Only when we invite, encourage, incentivize, and celebrate racial inclusion can we be confident that our data reflects the health needs of our population. This is important, because there are striking differences in disease susceptibility and risk among many of these groups.

How Can We Validate New Actionable Possibilities?

Just as we saw with systems biology, the theoretical benefits of scientific wellness will only become compelling when researchers are able to delineate clinically validated actionable possibilities. More extensive genome sequencing has paved the way to more targeted guidelines for prescribing actionable medications, and more individualized interventions. The same is beginning to happen with phenomic measurements such as blood proteins, blood metabolites, gut microorganisms, and other systems.

How Can We Persuade Payers to Assume the Cost of Wellness and Prevention?

The key to convincing government agencies and insurance companies that they should contribute to the costs associated with keeping people well longer, and not just treating disease, will be to demonstrate the enormous savings that will come from improving population health. Prior to the COVID-19 pandemic, as much as 86 percent of the $4 trillion spent on healthcare in the United States each year went to chronic diseases. If we can detect and reverse chronic diseases early, enormous savings will result. As a second example, only about 10 percent of people respond to the ten most commonly used drugs to treat various chronic conditions.[32] Phenome analyses will allow us to identify biomarkers that make it possible to distinguish responders from non-responders, potentially leading to savings of as much as 90 percent of the $600 billion we spend annually on drugs. Keeping individuals well is much less expensive than letting them become sick. In so many ways scientific wellness will lead to enormous cost savings. Payers eager to benefit from these savings will be motivated to act accordingly.

How Do We Decrease the Cost of Phenomic Analyses?

We predict that these costs will come down by as much as a hundredfold in the next ten years, but it will take new technological advances to reduce costs across the board. Savings will also be driven by the fact that inexpensive digital measurements will become ever more sensitive and broader in scope, generating myriad new actionable possibilities. Finally, as we assemble enormous amounts of data, we will be able to identify the most effective biomarkers and thus use perhaps a hundredfold fewer measurements to realize 90 percent of the actionable possibilities. All these advances will be driven by large-scale government-supported genome/longitudinal phenome programs. Our prediction is that this will eventually lead to a comprehensive at-home test that will make the approximately 5,000 measurements we need for a complete routine assessment of our health, uploaded to AI for analysis, resulting in the nearly immediate communication of actionable possibilities to our physicians and ourselves. Bringing scientific wellness to the home will lead to enormous savings in healthcare infrastructural costs now born by hospital- and clinic-based care.

How Will We Bring Wellness and Prevention to Healthcare Systems and Eventually to All Individuals?

Our experience at Arivale was that consumers loved the program and often left their data-skeptical physicians in search of more supportive doctors. We believe patients will be a strong driving force pushing healthcare systems (and reinforcing wellness companies) to adopt the science of wellness and prevention. We also believe that pioneering or champion physicians within healthcare systems will help lead the way to scientific wellness by recruiting their colleagues through increasing the quality of healthcare for their patients. Healthcare systems will follow once we have won the payers over to assuming the upfront wellness costs associated with the enormous downstream disease savings.

We also believe that existing wellness-centric practitioners or champions will begin to play a much larger role in academic centers for promoting scientific wellness. Community hospitals may be more venturesome at first than academic centers, which can often be paralyzed by academic skepticism. There are two other important groups of wellness advocates: the plethora of existing and emerging companies practicing various aspects of conventional and data-driven wellness, and the rapidly growing ranks of functional and personalized medicine physicians who embrace a holistic approach to health. Many of them now are embracing a data-rich approach to wellness.

How Do We Assure That This Form of Healthcare Is Available to Everyone?

Expanding access is clearly critical. There are massive efforts underway aimed at improving the metrics on diversity, equity, and inclusion, as all the million-person projects are attempting to do. This is a part of the mission now for most healthcare systems and large-scale research projects. To take an example we know well, Phenome Health has put together educational programs to convince patients, physicians, and other members of the healthcare community about the opportunities inherent in the science of wellness and prevention. Phenome Health is also partnering with groups that reach out to people from marginalized communities. Convincing them to join will be the prime challenge of the fourth P—participatory—persuading diverse patients, physicians, healthcare leaders, and regulators to actively join in the transformation from disease-oriented to wellness-oriented healthcare.

How Do We Solve the Four Major Challenges of Contemporary Healthcare: Poor Quality, Aging Populations, Exploding Chronic Diseases, and Unsustainable Cost Escalations?

The answers, as we hope we have convinced you, are to improve access to scientific wellness and make a concerted push for healthy aging, scanning for genetic risk for chronic diseases to treat high- and low-risk individuals differently, and following those who do have higher risks for the earliest signs of transitions followed by reversal. Healthy aging will also play a major role in delaying the onset of chronic diseases—as COVID-19 has so compellingly demonstrated. This also will require a lot of persuasion as we seek to convince people that they can—and should—play a far more active role in maximizing their own health potential.

Emergence of the Human Phenome Initiative

When Lee joined Providence in 2016, he began thinking about a genome/phenome effort that scaled up from the 5,000 individuals of Arivale. In 2021, Lee first proposed a follow-up study to the Human Genome Project to be funded by the government, scaling up 200-fold from Arivale—a proposed massive effort to sequence the genomes and determine the longitudinal phenomes of one million individuals over ten years. This is called the Human Phenome Initiative, a demonstration project that will validate the science of wellness and prevention and the many claims for data-driven health we have previously discussed. It will also provide multiple solutions to most of the seven challenges we discussed in the preceding section. Remember that the determinants of health delineate that lifestyle and environment account for 60 percent of health outcomes, while genes account for 30 percent (see Figure 11.2). Healthcare itself only accounts for 10 percent of the

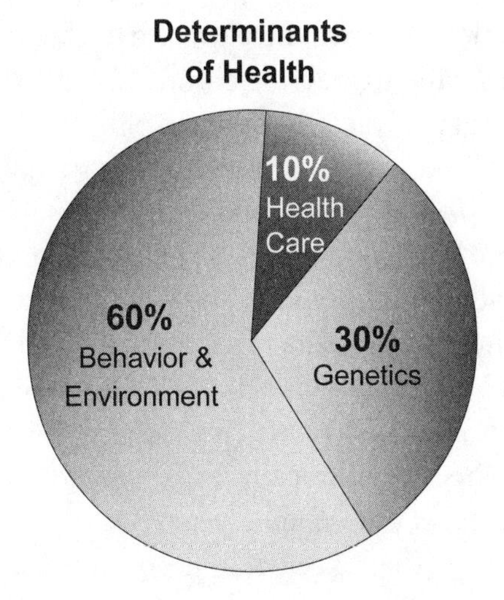

Figure 11.2. Only 10 percent of health outcomes can be attributed to medical interventions. While many people believe genetics are destiny, lifestyle choices and environmental context are a far greater determinant of health.

health outcomes—yet it absorbs almost all healthcare resources. Key components of lifestyle include diet, exercise, sleep, community, happiness, and stress. Important environmental inputs include toxins like black mold, smoke, heavy metal contaminants (lead and mercury, for example), and auto exhaust. Additional health-endangering contaminants include pesticides, hormone disruptors, smoke from wildfires, and other persistent health-deleterious chemicals. Blood and stool samples, collected every six months, would be placed in a biobank and would be available for data-driven hypothesis testing. The data, once collected, would be transferred to the cloud, standardized, analyzed, and integrated with different data types to create a totally unique health data ecosystem for discovery and innovation.

Phenome Health has partnered with Guardian Research Network (GRN), a nonprofit with access to 30 million patients in 120 different hospitals working across thirteen different states in the southern and southeastern United States. Guardian hospitals offer care to patients in community healthcare settings, including many Black, Latino, and low-income populations, making it possible to recruit participants for the Human Phenome Initiative that will better represent the racial diversity of the US population. Posit will help oversee the brain health objectives, and Technicity, a small engineering company, has built a software platform to manage and analyze the data. ISB will bring sophisticated new computational techniques to the effort, including a multiomic computational platform, knowledge graphs, and digital twins. Google will partner with us on search, cloud computing, digital health, and hopefully hyperscale AI (very powerful computation engines which will help us process extremely large and complex data sets from individual patients).

Several features distinguish the Human Phenome Initiative from its other million-person counterparts (including All of Us). First, it includes a highly effective, clinically corroborated digital approach to brain health. We expect to be able to predict common brain diseases such as depression, schizophrenia, and post-traumatic stress disorder with a high degree of accuracy and to be in a position to partially manage them early, before they become fully entrenched.

Second, the initiative will return thousands of newly clinically validated, actionable possibilities to patients and their physicians through the use of AI. These will enable powerful scientific wellness and a major commitment to healthy aging for each individual.

Third, it will be carrying out extensive longitudinal phenome analyses whereas most of the other million-person projects currently are not—and this will be essential to assessing the impacts of 60 percent of the determinants of health (e.g., lifestyle and environment).

Fourth, it will attempt to identify and eventually reverse the earliest wellness-to-disease transitions for individuals before they ever become clinical diseases. For example, we expect to see more than 200,000 wellness-to-disease transitions in this population over ten years and to validate markers for the early detection of most chronic diseases. This brings the possibilities of reversing many chronic diseases at an early stage.

Fifth, we will use the biological age metric to follow the wellness of individuals twice a year and use this to optimize healthy aging. This will delay the onset of chronic diseases for each individual—extending their period of wellness.

Sixth, the Human Phenome Initiative proposes to use the data-rich precision medicine approaches we tested in our studies of COVID-19 to attack the four major chronic diseases: type 2 diabetes, cardiovascular disease, cancer, and Alzheimer's disease. Our aim is to stratify these into different subtypes and to provide key insights into blood biomarkers, drug targets, and mechanisms of disease.

Seventh, the initiative will provide the computational, technical, educational, and clinical *infrastructure* for allowing other healthcare providers to join in this move toward the science of wellness and prevention. Finally, we are designing a variety of education programs to persuade patients, physicians, healthcare leaders and regulators, industrial partners, and others in the healthcare ecosystem to participate and join us in this revolution of healthcare.

Given the vast amount of data and discovery that came from just 5,000 Arivale clients, the potential inherent in one million genomes and phenomes, viewed in the context of their health profiles followed over up to ten years, is so vast as to be almost unimaginable. At a bare minimum, we believe that the project will be a powerful demonstration of what twenty-first century medicine can and should be. Moreover, it offers solutions to each of the five major challenges to healthcare—quality; an aging population; an explosion of chronic diseases; a lack of diversity, equity, and inclusion in healthcare data; and the ever-escalating costs of healthcare.

Pushing for Wellness

We believe patients, once educated, will be enthusiastic supporters of wellness. Over time we expect they will exert a powerful influence on the healthcare system by demanding that their doctors identify their specific predispositions with greater accuracy, monitor their progression, and help keep them healthy longer. Isolated "physician champions" in large healthcare systems, both academic and community-based, are starting to bring the tools of scientific wellness into their practices, but many of these are still struggling with the lack of cost incentives. Once large-scale human data-gathering projects of the types we have described show the enormous cost savings we have projected, payers will start to subsidize wellness and prevention thus accelerating change.

Across the United States and in parts of Europe, Asia, and the Middle East, one can find physicians who are setting up data-based wellness practices. In the United States, these include Wild Health, Cenegenics, Forward, Parsley Health, and many others. Functional medicine practitioners represent another highly networked group of physicians who take a systems view of disease and embrace the move toward holistic scientific wellness. Proponents of functional medicine constitute a powerful vector for change. As momentum builds, we believe they will expand rapidly in numbers given the right infrastructural support. Finally, there is a host of wellness companies, some promoting digital health approaches, providing dietary supplements, or enhancing microbiome-based gut health. Others are focusing on data integration and analysis. Nathan's efforts at Thorne seek to integrate many of these pieces to deliver personalized, scientific wellness to millions of people and through participating healthcare providers. All of these will be supported by the clinical validation of actionable possibilities, the demonstration of increased quality of healthcare, and the striking savings. In a sense the revolution will come from the bottom up (driven by patients and physicians), the top down (coming from hospitals and healthcare system leaders), and even the side (supported and embraced by advocates at wellness companies, proponents of integrative and functional medicine, and specialized practices).

We believe the Human Phenome Initiative and the other million-patient projects now well underway will provide such compelling data that large healthcare providers will have little choice but to begin transitioning their models of care to align with the principles of scientific wellness through actionable possibilities from both the genome and phenome and their integration. In some respects, the shift has already begun. Healthcare consulting groups such as Deloitte are now forecasting what we've been saying for years: that healthcare expenditure is shifting such that the active pursuit of wellness will be a bigger source of revenue than disease care by 2040.[33]

If large healthcare providers are to become part of the scientific wellness movement, they will have to be persuaded that they have a financial incentive to do so. In some cases, this is undeniable. But our healthcare system is not financially structured to keep us well; no one in the current system—other than the patient—benefits from extra years of health. Where P4 medicine helps drive immediate savings or earnings, there will be an opportunity to move these principles into the mainstream of healthcare. Where financial benefits are likely to come later, we will find continued resistance. Those of us who believe in these principles will have to be creative and innovative, meeting these interests head on and not in some theoretical ideal of how things should be or could be if health insurance companies did not determine the financial structures and constraints of American healthcare.

Your Role in the Revolution

There is another pathway that we believe to be essential to bringing this vision of wellness to life sooner than later: It must come from individuals who are not waiting for this kind of care to be embraced by their healthcare provider but are proactively seeking it for themselves. This can begin with an active attempt to understand the elements of scientific wellness as outlined in this book and think about how you would like to participate. It follows with an assertive search for doctors who have embraced elements of scientific wellness, both traditional and modest data-driven, in their practices. It is futile, at this stage, to hunt for those who are using huge genomic and phenomic datasets to guide their decision-making about individual patients. But there is a rapidly growing number of physicians who are open to using more personal data and analysis to guide their work with individual patients. These are the doctors who are excited about proactive and personalized approaches to maintaining wellness, who avoid the impulse to prescribe a drug until other options have been discussed, who understand that body health is nothing without brain health, and who embrace and can communicate the strategies that may slow or reverse the process of biological aging. These are doctors who will not brush aside a patient's inquiries about additional bloodwork or an exploration of their microbiome, who are eager to review measures of digital health, and who want to see—and, importantly, to measure—their patients when they are well, not just when they are sick.

These physicians exist, but they can be challenging to find. We have seen examples of such physicians at Providence, in small clinical practices, and among functional medicine physician practices. But the best advice we can offer is to begin the search as soon as possible, working through every doctor you have access to through your provider or other means, and helping those doctors understand why you're engaged in this journey.

A few words of warning are in order here: because P4 scientific wellness is not yet status quo medicine, there is no shortage of grifters, quacks, and snake-oil salesmen who are marketing themselves as wellness specialists but aren't interested in or able to understand actual data-driven science. Patients need to be smart and vigilant, particularly when dealing with anyone who is not a licensed physician who wants to use measurements that are beyond the scope of those discussed in this book, or who recommend therapies that have not been clinically validated and found to be safe. It is important to remember that there is a big difference between rejecting a system that is focused on diseases rather than wellness, and rejecting actual science.

If you want to take your health into your own hands and are wondering where to start, there are a number of tests available for generating information to inform your wellness journey. Beyond the standard lipid panels and measurements of fasting glucose in a comprehensive metabolic panel or complete blood cell count, these include blood tests for genome and phenome data, stool tests for the gut microbiome, and saliva tests for stress hormones. Digital devices measure heart and blood pressure, blood oxygen and glucose levels, activity, quality of sleep, heart rate variability (to assess your balance between fight/flight and calm—the sympathetic and parasympathetic nervous systems, respectively), and more generally promote body and brain health. It is always beneficial to find a physician who can and will help you interpret these results.

Over the past few years, a growing number of doctors have turned their attention to helping their patients obtain accurate measures of biological aging. Wellness-oriented physicians are likely to agree that an ongoing assessment of biological age, year after year, can help track and optimize healthy aging. One important caveat is that it is becoming clear that our bodies do not age in a uniform way. Our brains often age at a different rate than our hearts, our immune systems, or other organs, tissues, and systems. So while an overall biological aging metric is a valuable way to understand how we have aged holistically, physicians and companies that can help individuals see how they are aging at a molecular level in each of their major organs, while suggesting possible approaches to optimize individual organ aging, are several steps ahead of the wellness curve.

Additionally, there are several digital approaches to measuring your physiology that are reliable and reasonably priced, and that a credible wellness-oriented doctor will likely find useful. These include wearable devices that track one's activity, body temperature, cardiovascular measurements, respiratory measurements, and sleep quality, as well as some devices under development that offer real-time tracking not only of blood sugar but also of the electrolytes and metabolites in perspiration and even blood pressure.[34]

Credible wellness-centric physicians should believe they have a responsibility to ensure that their patients have the information and tools they need to maintain neural plasticity and effective cognitive functioning. This absolutely must include regular brain health assessments—even exercise is far less effective and far less efficient when we do not have a handle on what needs to be exercised. Doctors shouldn't wait for patients to complain of trouble with memory or focus before offering these assessments; brain health should be part of the regular routine of the annual physical from the age of 50 on, if not sooner. This can be achieved through testing provided by Posit Science (BrainHQ), Cambridge Brain Sciences, or CNS Vital Signs. These efforts should be facilitated with tools that have been evaluated in a research setting and shown to be effective at improving cognitive fitness.

No physician who truly believes in scientific wellness will be dismissive of a patient who wishes to understand the research behind a test, treatment, or therapy. If you ever find yourself working with a doctor who scoffs at such a request, or who suggests that their techniques are so "cutting edge" that "the research hasn't caught up," you would be smart to search for care elsewhere.

Not long ago it was hard to find a doctor who was enthusiastic about the elements of scientific wellness. Today, a growing number of physicians are not content to care for their patients based on what they learned back in medical school, recognizing that "the way we've always done it" is an out-of-date and out-of-touch approach for the prevention, diagnosis, and treatment of disease. Doctors who have not yet had this awakening are not bad people. Most physicians care deeply about their patients and there are virtually none who wish for the people in their care to be sick, in pain, or otherwise suffering. But if your relationship with your primary care physician is one that is predicated on having symptoms before you can schedule an appointment, it might be time for a change. And even if you are one of the very small number of people who gets the currently "recommended" preventative care—just 8 percent of all US adults over the age of 35, according to a recent study—it is important to recognize that those services are almost exclusively aimed at looking for classic symptoms of disease, like cholesterol and high blood pressure.[35] And, as we now know, if you want a lifetime of good health it is imperative that you not wait for disease to become symptomatic before addressing the issue.

If you haven't been having regular blood tests, why not? If those tests are done, how will they be monitored? Will they be evaluated with an eye toward your genetic risks? If so, how will those risks be ascertained? What are your doctors doing to measure and monitor your brain health? Have they ever mentioned the importance of your gut microbiome?

Any doctor who won't entertain these questions should be left as soon as a better physician can be found. Any doctor who will talk about these matters but not take action should be left as soon as a better physician can be found. Any doctor who will offer these services so they can bill you, but not engage in the work of using these practices to help you stay well, should be left as soon as a better physician can be found. That might sound harsh, but consider what is at stake: the only life you've got.

This won't just be good for you. It will help push doctors and the providers they work with to understand the importance offering scientific wellness for everyone. This is how revolutions are born.

The Steps You Can Take Right Now

Even without a doctor at your side, there is plenty that you can do to apply the principles of scientific wellness in your own life.

The easiest and least expensive way to get started is by tracking your own measures of digital health and using these to help improve habits related to wellness pillars such as exercise, diet, sleep, and stress management. Today's top-selling activity trackers will count your steps, level of vigorous activity, pulse, amount of sleep, calories burned and more, and allow for easy storage and access of past data. More advanced wearable technology offers not just metrics of activity but also respiration, temperature, sophisticated measures of heart rate variability, and stages of sleep. If you have a family history of diabetes, you might consider wearing at regular intervals a continuous glucose monitor to determine how your diet and activity affect your blood glucose levels. Maintaining safe blood glucose levels is one of the most important aspects of wellness you can control.

For those who are up for it, coupling that data with a running log of the food you eat for a period of time offers exponentially greater insights into your well-being. This isn't as easy as slapping a watch on your wrist, but it isn't much more difficult than that, either—and it costs almost nothing. Free and inexpensive food tracking apps like MyFitnessPal make the act of tallying calories and nutrients as easy as sending a text message, and take no more time than that. Doing this for a week or two can be really eye opening and helpful for a better understanding where your nutrition stands. Maintaining a proper body weight is another critical element of health.

The data you get from these two sources is more valuable if you understand the raw material you're working with. Just about everyone can benefit from whole genome sequencing. This is more expensive than the limited personal genetic tests that have become quite cheap and very popular, like those aimed at helping people build family histories, but there is tremendous value in having insights into your whole genome and its implications for wellness and disease, and many people will find that the additional investment is worthwhile. Fortunately, the costs for whole genome sequencing are coming down quickly as competition builds between companies that are making this service available, such as Gencove, Nebula Genomics, or several new companies coming out with novel single-molecule sequencing technologies. Some of these predict that the cost of whole genomes will drop to $10 in the not-too-distant future. Most of these companies will provide a report that highlights at least some of your genetic risks and insights into your unique physiology, but these are almost never comprehensive. Scientists are discovering more genome correlations with wellness or disease every day, so what you learn today about your genetic propensity for disease is likely to be only a fraction of what will be known in another five-to-ten years' time. For this reason, it is advisable to work with a company that has a system in place to update you when new actionable possibilities are discovered.

One of the very big advantages of whole genome sequencing is its ability to help you decide what biomarkers to monitor because for most people, right now, it would be both prohibitively expensive and overwhelmingly complicated to test for every potentially meaningful marker. Increasingly, commercial genomic test results come with recommendations for the markers of greatest potential interest, permitting you to choose from some of the most valuable and easy-to-understand tests. The genome can provide insight into your disease risk profiles, and which conditions you may want to test for most often and most carefully. These tests, if not ordered by a doctor or covered by insurance, can be obtained from a variety of places, including Quest Diagnostics, Labcorp, LetsGetChecked, and Thorne.

Another word of warning: it's easy to get carried away, to become myopic about the meaning of a single test, and to forget that test results offer a snapshot of one small aspect of your body at one particular time, during a blood draw or stool swab that takes seconds to complete. Like our global climate, our bodies are subject to significant variability, and it is only by keeping careful watch over time that we can better understand what is happening to our individual systems in the context of our lifestyles and environmental exposures.

For this reason, it's helpful to have a team of experts in your corner. That's what we were able to provide for the thousands of people enrolled in Arivale and what participants in the Human Phenome Initiative will have access to. It is what boutique healthcare providers like Wild Health and Parsley Health offer. It's what lots of functional medicine physicians around the country are now providing. And it is a standard of care that we believe will become widely available in the near future as the price point for such services continues to fall and new actionable possibilities are validated from genome and phenome data.

Big Questions on the Cusp of Answers

We believe that in the coming decade people in the United States, western Europe, and parts of Asia will have increasing access to medical care that embraces data-driven wellness. Genetic screening and drawing more extensive and regular information from the blood and biomes, when converted into actionable possibilities, will fundamentally alter their life trajectories. This will already be an exciting development, but what's coming just a little further down the line will be absolutely stunning. For we have only probed the surface of understanding the single bad genes, paired bad genes, and polygenetic risk scores that will help us navigate our personal health journey when it comes to diseases, injuries, neurological health, biological aging, immune system health, pharmaceuticals, supplements, food, exercise, stress, diet, and other lifestyle factors.

Even less explored right now, but potentially more profoundly impactful in the long-term, will be the thousands of new actionable possibilities offered by a generation of dense longitudinal data clouds, and by the integration of genomic and phenomic information with data on the gut microbiome, electronic health records, patient reported outcomes, and determinants of health. As the complexity will surpass any human's ability to fully understand, artificial intelligence will be needed to deliver relevant possibilities for each individual to their physician in a prioritized manner, with simple explanations of what each actionable possibility represents. These features are exactly what the Human Phenome Initiative is pioneering.

Did it work? Did it fail? How fast did it happen? What systems were affected? What were the outcomes after one month? And one year? And one decade? Each answer will become new data. Outliers will be of exceptional value, providing mechanistic insight into wellness and disease. If a person whose genetics seem to suggest they should have a disease doesn't get it, why? If a person whose disease trajectory seems to suggest that a particular intervention should work but it doesn't, why not?

At 100 people, like those who enrolled in the Pioneer 100 project, we're able to point to likely solutions for a few common problems. At 1,000 individuals, we're flooded with insights to improve outcomes as with Arivale. At one million persons, even the most exceptional rare disease cases will have generated actionable data.

Answers that might only have come when researchers could fund exceptionally expensive longitudinal studies of tens of thousands of individuals will now be much more accessible. The fact that multidimensional longitudinal phenomic data can now be generated on all patients means that far fewer patients will be needed in clinical trials to obtain highly informative results. Some researchers are already experimenting with two-step clinical trials, where each step may only use about 100 patients.[36] With genomic and longitudinal phenomic analyses in the first step, we can, for example, identify a panel of blood biomarkers that will distinguish drug responders from non-responders.[37]

In the second step, we can use 100 predicted responder patients to actually test the drug. If more than 90 percent of these patients respond, then it is likely that the FDA would approve the drug as it did with 44 of 46 patients successfully treated with Herceptin, a breast cancer drug.

This approach could enormously reduce the cost of drug testing—and do a lot to speed that testing up. And if there's something that became crystal clear to us during the COVID-19 pandemic it is that faster identification of effective therapies can be a matter of life and death.

In the long run, our COVID-19 observational study demonstrated that, with genome and phenome analysis, and deep immune phenotyping, scientists can draw powerful conclusions from a small number of patients. Well-designed studies of a few hundred people may no longer need to be considered "small," "incomplete," and "inconclusive." We showed that we can identify a disease we want to understand, investigate its trajectory in a few hundred patients with the tools of genotyping and deep phenotyping, and emerge with a wealth of knowledge about its trajectory, giving us a chance to quickly identify avenues for early intervention.

The Future We Expect

COVID helped educate the public about the impact of their lifestyle decisions on their health outcomes. Over the past few years, for a variety of intersecting reasons, more physicians are bringing the concepts of scientific wellness into their practices every day. More people are beginning to collect and track their health data, and to change their behavior accordingly.

In many respects, however, we still have a long way to go before we can fully realize the possibilities of scientific wellness. A healthcare system that was designed to combat infectious diseases still dominates medicine as it is practiced today. And there are a great many special interests who profit handsomely from the status quo. If those forces have their way, this vision won't be twenty-first-century healthcare, it will be twenty-second-century healthcare.

But research scientists won't be comfortable sitting on this kind of potential for too long without acting on it. Sooner or later, we'll reach a realm of predictive, preventive, personalized, and participatory care. If the Human Phenome Initiative and other large-scale population programs are successful, they could have a profound impact, propelling various actors working in our complex healthcare system to embrace the science of wellness and prevention.

If we want to get there sooner rather than later, there's one change we can make—indeed, that we can demand—that could make a big difference. It's called "value-based healthcare," a compensation program that doesn't pay doctors based on how many patients they see, how many procedures they perform, or how many tests they run, but on how many patients' lives they meaningfully improve. With the cost savings that will come from the ten-year demonstration of the Human Phenome Initiative and its many national and international counterparts, payers and providers will gradually buy into scientific wellness on economic as well as clinical and scientific grounds. Once this happens, value-based healthcare will become one of many consequences of the seismic shift from a disease-oriented to a wellness-oriented healthcare system.

Another book would have to be written to do justice to value-based healthcare—and thankfully several have been—but in short this is a system that values quality rather than quantity, and wellness rather than disease. We believe the huge savings that will come from the Human Phenome Initiative and the other million-person studies underway around the world today will be fundamental in catalyzing this revolution. If healthcare systems and physicians are paid to keep their patients healthy, instead of treating them (and earning money) only when they are sick, then we will finally have an economic driver for wellness and prevention. Scientific wellness and healthy aging will be the most efficient means to achieve this objective.

Does this sound like a tall order? Absolutely. But even a system with a strong financial incentive to maintain the status quo will eventually see the quality and cost savings that will come from this new science. In countries with single payer systems, the financial incentives for wellness are more easily aligned, given the shared interest in maintaining health.

So what is the healthcare future you wish to see? Is it predictive, preventive, personalized, and participatory? Is it based on scientific wellness, healthy aging, the use of genetics to optimize wellness, the early reversal of wellness-to-disease transitions, and the use of data-rich observational studies to understand the nature and trajectory of chronic disease? Does it recognize that brain health is every bit as important as body health? Is it predicated on the notion that doctors and healthcare systems should only do well when patients live well? Does it envision a world in which cancer, heart disease, Alzheimer's, diabetes, autoimmune conditions, and other chronic diseases could be preventable? Does it offer a future in which you will live in wellness and maintain your vitality, creativity, energy, passions, physicality, intelligence, and community connections into your nineties or 100s?

If that's the future you want, then we have some good news for you: while this might have been unthinkable just a few years ago, we believe scientific wellness is on its way. The vital question is: When will it arrive? We cannot fully answer this yet. But we can say with confidence that every one of us can play a role in bringing this vision to fruition sooner than later. It starts with knowing what is possible and understanding that, for every one of us, our future health is in our own hands.

Notes

Acknowledgments

Index

Notes

Introduction

1. M. Xue et al., “Diabetes Mellitus and Risks of Cognitive Impairment and Dementia: A Systematic Review and Meta-Analysis of 144 Prospective Studies,” *Ageing Research Reviews* 55 (2019): 100944; D. Glovaci, W. Fan, and N. D. Wong, “Epidemiology of Diabetes Mellitus and Cardiovascular Disease,” *Current Cardiology Reports* 21 (2019): 21.

2. L. Junjun et al., “Prognosis and Risk Factors in Older Patients with Lung Cancer and Pulmonary Embolism: A Propensity Score Matching Analysis,” *Scientific Reports* 10 (2020): 1272.

3. P. Boersma, L. I. Black, and B. W. Ward, “Prevalence of Multiple Chronic Conditions among US Adults,” *Preventing Chronic Disease 2020 (CDC Report)* 17 (2018): 200130.

4. J. S. Paulsen et al., “Detection of Huntington’s Disease Decades before Diagnosis: The Predict-HD Study,” *Journal of Neurology, Neurosurgery, and Psychiatry* 79 (2008): 874–880; M. Gerstung, “The Evolutionary History of 2,658 Cancers,” *Nature* 578 (2020): 122–128; A. T. Magis et al., “Untargeted Longitudinal Analysis of a Wellness Cohort Identifies Markers of Metastatic Cancer Years Prior to Diagnosis,” *Scientific Reports* 10 (2020): 16275; K. B. Rajan et al., “Cognitive Impairment 18 Years before Clinical Diagnosis of Alzheimer Disease Dementia,” *Neurology* 85 (2015): 898–904.

5. B. Gallo Marin et al., “Predictors of COVID-19 Severity: A Literature Review,” *Reviews in Medical Virology* 31 (2021): 1–10.

6. Centers for Disease Control and Prevention, “Severe Outcomes among Patients with Coronavirus Disease 2019 (COVID-19),” *Morbidity and Mortality Weekly Report* 69 (2020): 343–346.

7. N. J. Schork, “Personalized Medicine: Time for One-Person Trials,” *Nature* 520 (2015): 609–611.

8. L. Timmerman, *Hood: Trailblazer of the Genomic Age* (Bandera Press, 2016).

9. R. M. Hewick et al., "A Gas-Liquid Solid Phase Peptide and Protein Sequenator," *Journal of Biological Chemistry* 256 (1981): 7990–7997; L. M. Smith et al., "Fluorescence Detection in Automated DNA Sequence Analysis," *Nature* 321 (1986): 674–679; S. J. Horvath et al., "An Automated DNA Synthesizer Employing Deoxynucleoside 3'-Phosphoramidites," *Methods in Enzymology* 154 (1987): 314–326; G. K. Geiss et al., "Direct Multiplexed Measurement of Gene Expression with Color-Coded Probe Pairs," *Nature Biotechnology* 26 (2008): 317–325; Z. Guo, L. Hood, and E. W. Petersdorf, "Oligonucleotide Arrays for High Resolution HLA Typing," *Reviews in Immunogenetics* 1 (1999): 220–230; M. Hunkapiller et al., "A Microchemical Facility for the Analysis and Synthesis of Genes and Proteins," *Nature* 310 (1984): 105–111.

10. R. Dulbecco, "A Turning Point in Cancer Research: Sequencing the Human Genome," *Science* 231 (1986): 1055–1056.

11. F. Dyson, *Imagined Worlds* (Harvard University Press, 1997).

12. L. Hood and L. Rowen, "The Human Genome Project: Big Science Transforms Biology and Medicine," *Genome Medicine* 5 (2013): 79.

13. Hood and Rowen, "The Human Genome Project," 79.

14. T. Ideker, T. Galitski, and L. Hood, "A New Approach to Decoding Life: Systems Biology," *Annual Review of Genomics and Human Genetics* 2 (2001): 343–372.

15. Y. Lazebnik, "Can a Biologist Fix a Radio?—Or, What I Learned While Studying Apoptosis," *Cancer Cell* 2, no. 3 (2002): 179–182.

16. Ideker, Galitski, and Hood, "A New Approach to Decoding Life: Systems Biology," 343–372.

17. L. Hood et al., "Systems Biology at the Institute for Systems Biology," *Briefings in Functional Genomics and Proteomics* 7 (2008): 239–248.

18. T. Ideker et al., "Integrated Genomic and Proteomic Analyses of a Systematically Perturbed Metabolic Network," *Science* 292 (2001): 929–934.

19. R. Bonneau et al., "A Predictive Model for Transcriptional Control of Physiology in a Free Living Cell," *Cell* 131 (2007): 1354–1365.

20. D. Hwang et al., "A Systems Approach to Prion Disease," *Molecular Systems Biology* 5 (2009): 252.

21. National Research Council, *A New Biology for the 21st Century* (The National Academies Press, 2009), https://doi.org/10.17226/12764.

22. M. Berretta et al., "Physician Attitudes and Perceptions of Complementary and Alternative Medicine (CAM): A Multicentre Italian Study," *Frontiers in Oncology* 10 (2020): 594.

23. M. Tai-Seale, T. G. McGuire, and W. Zhang, "Time Allocation in Primary Care Office Visits," *Health Services Research* 42 (2007): 1871–1894.

24. N. D. Price et al., "A Wellness Study of 108 Individuals Using Personal, Dense, Dynamic Data Clouds," *Nature Biotechnology* 35 (2017): 747–756.

1. An Infectious Idea

1. Centers for Disease Control and Prevention, "Control of Infectious Diseases," *The Morbidity and Mortality Weekly Report* 48 (1999): 621–629.

2. T. N. Bonner, "Searching for Abraham Flexner," *Academic Medicine* 73 (1998): 160–166.

3. A. Flexner, *I Remember: The Autobiography of Abraham Flexner* (Simon and Schuster, 1940).

4. Flexner, *I Remember.*

5. G. M. Bartelds et al., "Development of Antidrug Antibodies against Adalimumab and Association with Disease Activity and Treatment Failure during Long-Term Follow-Up," *Journal of the American Medical Association* 305 (2011): 1460–1468.

6. Alzheimer's Association, "Alzheimer's Disease Treatment Horizons," updated October 2019, https://www.alz.org/media/homeoffice/teaser%20image/alzheimers-dementia-disease-treatment-horizons-ts.pdf.

7. S. Leucht et al., "How Effective Are Common Medications: A Perspective Based on Meta-Analyses of Major Drugs," *BMC Medicine* 13 (2015): 253.

8. Leucht et al., "How Effective Are Common Medications," 253.

9. N. J. Schork, "Personalized Medicine: Time for One-Person Trials," *Nature* 520 (2015): 609–611.

10. A. B. Martin et al., "National Health Care Spending in 2017: Growth Slows to Post–Great Recession Rates; Share of GDP Stabilizes," *Health Affairs* 38 (2019): 96–106.

11. V. Raghupathi and W. Raghupathi, "Healthcare Expenditure and Economic Performance: Insights from the United States Data," *Frontiers in Public Health* 8 (2020): 156.

12. N. K. Mehta, L. R. Abrams, and M. Myrskyla, "US Life Expectancy Stalls Due to Cardiovascular Disease, Not Drug Deaths," *Proceedings of the National Academy of Sciences of the United States of America* 117 (2020): 6998–7000.

13. S. L. Murphy et al., *Mortality in the United States, 2020* (National Center for Health Statistics Data Brief No. 427, 2021).

14. K. D. Kochanek, R. N. Anderson, and E. Arias, "Changes in Life Expectancy at Birth, 2010–2018" (NCHS Health E-Stat, 2020).

15. Centers for Disease Control and Prevention, "Percent of U.S. Adults 55 and Over with Chronic Conditions," last reviewed November 6, 2015, https://www.cdc.gov/nchs/health_policy/adult_chronic_conditions.htm.

16. P. Kudesia et al., "The Incidence of Multimorbidity and Patterns in Accumulation of Chronic Conditions: A Systematic Review," *Journal of Multimorbidity and Comorbidity* 11 (2021).

17. "Don't Let the Good Life Pass You By," dir. Michael Schur, *The Good Place*, season 3, ep. 8, aired November 15, 2018.

18. "FAIR Health Consumer," https://www.fairhealthconsumer.org/.

19. J. Wapner, "COVID-19: Medical Expenses Leave Many Americans Deep in Debt," *The BMJ* 370 (2020): m3097.

20. W. M. Sage, "Fracking Health Care: The Need to De-Medicalize America and Recover Trapped Value for Its People," (draft, May 31, 2017), https://petrieflom.law.harvard.edu/assets/publications/Sage_Fracking_Health_Care.pdf.

21. M. Rae et al., "Long-Term Trends in Employer-Based Coverage," Peterson-KFF Health System Tracker, April 3, 2020, https://www.healthsystemtracker.org/brief/long-term-trends-in-employer-based-coverage/.

22. K. M. Adams, W. S. Butsch, and M. Kohlmeier, "The State of Nutrition Education at US Medical Schools," *Journal of Biomedical Education* (2015): 357627.

23. *PBS News Hour*, May 8, 2017, https://www.pbs.org/newshour/show/improve-patient-diets-doctor-kitchen.

24. B. J. Cardinal et al., "If Exercise Is Medicine, Where Is Exercise in Medicine? Review of U.S. Medical Education Curricula for Physical Activity–Related Content," *Journal of Physical Activity and Health* 12 (2015): 1336–1343; J. A. Mindell et al., "Sleep Education in Medical School Curriculum: A Glimpse across Countries," *Sleep Medicine* 12 (2011): 928–931; A. Nerurkar et al., "When Physicians Counsel about Stress: Results of a National Study," *JAMA Internal Medicine* 173 (2013): 76–77.

25. A. Bradford et al., "Missed and Delayed Diagnosis of Dementia in Primary Care: Prevalence and Contributing Factors," *Alzheimer Disease and Associated Disorders* 23 (2009): 306–314.

26. I. England, D. Stewart, and S. Walker, "Information Technology Adoption in Health Care: When Organisations and Technology Collide," *Australian Health Review* 23 (2000): 176–185.

27. "Physician Salary Report 2022: Physician Income Rising Again," Weatherby Healthcare, May 19, 2022, https://weatherbyhealthcare.com/blog/annual-physician-salary-report.

28. M. Treskova-Schwarzbach et al., "Pre-Existing Health Conditions and Severe COVID-19 Outcomes: An Umbrella Review Approach and Meta-Analysis of Global Evidence," *BMC Medicine* 19 (2021): 212; J. J. Zhang et al., "Risk and Protective Factors for COVID-19 Morbidity, Severity, and Mortality," *Clinical Reviews in Allergy & Immunology* (2022): 1–18; A. Ramaswamy et al., "Patient Satisfaction with Telemedicine during the COVID-19 Pandemic: Retrospective Cohort Study," *Journal of Medical Internet Research* 22 (2020): e20786.

29. A. F. Bryan and T. C. Tsai, "Health Insurance Profitability during the COVID-19 Pandemic," *Annals of Surgery* 273 (2021): e88–e90.

30. Bryan and Tsai, "Health Insurance Profitability during the COVID-19 Pandemic," e88–e90.

31. C. Stewart, "Total Amount of Global Healthcare Data Generated in 2013 and a Projection for 2020," Statista, September 24, 2020, https://www.statista.com/statistics/1037970/global-healthcare-data-volume/.

32. M. D. Aldridge and A. S. Kelley, "The Myth Regarding the High Cost of End-of-Life Care," *American Journal of Public Health* 105 (2015): 2411–2415.

33. N. Zubair et al., "Genetic Predisposition Impacts Clinical Changes in a Lifestyle Coaching Program," *Scientific Reports* 9 (2019): 6805; M. Wainberg et al., "Multiomic Blood Correlates of Genetic Risk Identify Presymptomatic Disease Alterations," *Proceedings of the National Academy of Sciences of the United States of America* 117 (2020): 21813–21820.

34. On He Jiankui, see D. Normile, "Chinese Scientist Who Produced Genetically Altered Babies Sentenced to 3 Years in Jail," *Science*, December 30, 2019. J. Zaritsky, dir., *Do You Really Want to Know?* Optic Nerve Films, 2012, http://www.doyoureallywanttoknowfilm.com.

35. Priscilla Chan, interview, "Priscilla Chan on Husband Mark Zuckerberg's 'Totally Different Mentality,'" interview by Norah O'Donnell, *CBS Mornings*, February 19, 2019.

36. I. Johnston, "World Cancer Day 2017: Effective Cure Will Happen in Five to 10 Years, Says Leading Expert," *Independent*, February 2, 2017, https://www.independent.co.uk/life-style/health-and-families/health-news/world-cancer-day-2017-effective-cure-will-happen-five-to-10-years-expert-karol-sikora-a7558846.html.

37. M. V. Blagosklonny, "The Goal of Geroscience Is Life Extension," *Oncotarget* 12 (2021): 131–144.

38. Y. Cao et al., "Transplantation of Chondrocytes Utilizing a Polymer-Cell Construct to Produce Tissue-Engineered Cartilage in the Shape of a Human Ear," *Plastic and Reconstructive Surgery* 100 (1997): 297–302; discussion 303–304.

39. G. Zhou et al., "In Vitro Regeneration of Patient-Specific Ear-Shaped Cartilage and Its First Clinical Application for Auricular Reconstruction," *eBio-Medicine* 28 (2018): 287–302.

40. T. Sahakyants and J. P. Vacanti, "Tissue Engineering: From the Bedside to the Bench and Back to the Bedside," *Pediatric Surgery International* 36 (2020): 1123–1133.

41. J. Washington, "New Poll Shows Black Americans Put Far Less Trust in Doctors and Hospitals Than White People," Andscape, https://andscape.com/features/new-poll-shows-black-americans-put-far-less-trust-in-doctors-and-hospitals-than-white-people/.

42. D. U. Himmelstein et al., "Medical Bankruptcy: Still Common Despite the Affordable Care Act," *American Journal of Public Health* 109 (2019): 431–433.

2. Catalyzing the Revolution

1. J. C. Roach et al., "Analysis of Genetic Inheritance in a Family Quartet by Whole-Genome Sequencing," *Science* 328 (2010): 636–639.

2. M. Y. Brusniak et al., "ATAQS: A Computational Software Tool for High Throughput Transition Optimization and Validation for Selected Reaction Monitoring Mass Spectrometry," *BMC Bioinformatics* 12 (2011): 78; T. Farrah et al., "A High-Confidence Human Plasma Proteome Reference Set with Estimated Concentrations in Peptideatlas," *Molecular & Cellular Proteomics* 10 (2011): M110.006353.

3. G. K. Geiss et al., "Direct Multiplexed Measurement of Gene Expression with Color-Coded Probe Pairs," *Nature Biotechnology* 26 (2008): 317–325.

4. C. Lausted et al., "POSaM: A Fast, Flexible, Open-Source, Inkjet Oligonucleotide Synthesizer and Microarrayer," *Genome Biology* 5 (2004): R58.

5. H. D. Agnew et al., "Iterative in Situ Click Chemistry Creates Antibody-Like Protein-Capture Agents," *Angewandte Chemie International Edition English* 48 (2009): 4944–4948.

6. D. Hwang et al., "A Systems Approach to Prion Disease," *Molecular Systems Biology* 5 (2009): 252.

7. J. M. Bockman et al., "Creutzfeldt-Jakob Disease Prion Proteins in Human Brains," *New England Journal of Medicine* 312 (1985): 73–78; N. Jankovska, R. Matej, and T. Olejar, "Extracellular Prion Protein Aggregates in Nine Gerstmann-Sträussler-Scheinker Syndrome Subjects with Mutation P102l: A Micromorphological Study and Comparison with Literature Data," *International Journal of Molecular Sciences* 22 (2021); L. Baldelli and F. Pro-

vini, "Fatal Familial Insomnia and Agrypnia Excitata: Autonomic Dysfunctions and Pathophysiological Implications," *Autonomic Neuroscience* 218 (2019): 68–86.

8. S. A. Ament et al., "Rare Variants in Neuronal Excitability Genes Influence Risk for Bipolar Disorder," *Proceedings of the National Academy of Sciences of the United States of America* 112 (2015): 3576–3581.

9. K. R. Merikangas et al., "Lifetime and 12-Month Prevalence of Bipolar Spectrum Disorder in the National Comorbidity Survey Replication," *Archives of General Psychiatry* 64 (2007): 543–552.

10. F. J. A. Gordovez and F. J. McMahon, "The Genetics of Bipolar Disorder," *Molecular Psychiatry* 25 (2020): 544–559.

11. S. S. Mahmood et al., "The Framingham Heart Study and the Epidemiology of Cardiovascular Disease: A Historical Perspective," *Lancet* 383 (2014): 999–1008.

12. R. Chen et al., "Personal Omics Profiling Reveals Dynamic Molecular and Medical Phenotypes," *Cell* 148 (2012): 1293–1307.

13. L. Smarr, "Quantifying Your Body: A How-To Guide from a Systems Biology Perspective," *Biotechnology Journal* 7 (2012): 980–991.

14. M. Bowden, "The Measured Man," *The Atlantic*, July/August 2012, https://www.theatlantic.com/magazine/archive/2012/07/the-measured-man/309018/.

15. N. D. Price et al., "A Wellness Study of 108 Individuals Using Personal, Dense, Dynamic Data Clouds," *Nature Biotechnology* 35 (2017): 747–756.

16. O. Manor et al., "A Multi-Omic Association Study of Trimethylamine N-Oxide," *Cell Reports* 24 (2018): 935–946.

17. R. Dresser, "Subversive Subjects: Rule-Breaking and Deception in Clinical Trials," *Journal of Law, Medicine & Ethics* 41 (2013): 829–840.

18. Institute of Medicine, (US) Committee on Policies for Allocating Health Sciences Research Funds, *Funding Health Sciences Research: A Strategy to Restore Balance* (Washington, DC: National Academies Press, 1990), 2.

19. S. R. Chekroud et al., "Association between Physical Exercise and Mental Health in 1.2 Million Individuals in the USA between 2011 and 2015: A Cross-Sectional Study," *Lancet Psychiatry* 5 (2018): 739–746.

20. T. Mann et al., "Medicare's Search for Effective Obesity Treatments: Diets Are Not the Answer," *American Psychologist* 62 (2007): 220–233.

21. A. T. Skuladottir et al., "A Meta-Analysis Uncovers the First Sequence Variant Conferring Risk of Bell's Palsy," *Scientific Reports* 11 (2021): 4188.

22. V. Bhatia and R. K. Tandon, "Stress and the Gastrointestinal Tract," *Journal of Gastroenterology and Hepatology* 20 (2005): 332–339.

23. A. S. O. Yu et al., "Electric Vehicles: Struggles in Creating a Market," *Proceedings of Portland International Conference on Management of Engineering and Technology (PICMET) '11: Technology Management in the Energy Smart World* (2011): 1–13.

3. Mining Vast Troves of Information

1. D. White and M. Rabago-Smith, "Genotype-Phenotype Associations and Human Eye Color," *Journal of Human Genetics* 56 (2011): 5–7.

2. B. A. Moore et al., "Identification of Genes Required for Eye Development by High-sThroughput Screening of Mouse Knockouts," *Communications Biology* 1 (2018): 236.

3. S. P. Claus, H. Guillou, and S. Ellero-Simatos, "The Gut Microbiota: A Major Player in the Toxicity of Environmental Pollutants?" *NPJ Biofilms and Microbiomes* 2 (2016): 16003.

4. A. L. Richards et al., "Gut Microbiota Has a Widespread and Modifiable Effect on Host Gene Regulation," *mSystems* 4 (2019).

5. S. Mendis, "The Contribution of the Framingham Heart Study to the Prevention of Cardiovascular Disease: A Global Perspective," *Progress in Cardiovascular Diseases* 53 (2010): 10–14.

6. M. Gerstung et al., "The Evolutionary History of 2,658 Cancers," *Nature* 578 (2020): 122–128.

7. G. W. Small et al., "Cerebral Metabolic and Cognitive Decline in Persons at Genetic Risk for Alzheimer's Disease," *Proceedings of the National Academy of Sciences of the United States of America* 97 (2000): 6037–6042; G. Chetelat et al., "Amyloid-PET and (18)F-FDG-PET in the Diagnostic Investigation of Alzheimer's Disease and Other Dementias," *Lancet Neurology* 19 (2020): 951–962.

8. M. S. Shoukry et al., "The Emerging Role of Circulating Tumor DNA in the Management of Breast Cancer," *Cancers* 13 (2021); D. Crosby, "Delivering on the Promise of Early Detection with Liquid Biopsies," *British Journal of Cancer* 126 (2022): 313–315; N. Vidula, L. W. Ellisen, and A. Bardia, "Clinical Application of Liquid Biopsies to Detect Somatic BRCA1/2 Mutations and Guide Potential Therapeutic Intervention for Patients with Metastatic Breast Cancer," *Oncotarget* 12 (2021): 63–65.

9. A. R. Martin et al., "Clinical Use of Current Polygenic Risk Scores May Exacerbate Health Disparities," *Nature Genetics* 51 (2019): 584–591.

10. N. S. Guest et al., "Sport Nutrigenomics: Personalized Nutrition for Athletic Performance," *Frontiers in Nutrition* 6 (2019): 8.

11. G. T. Goodlin et al., "Applying Personal Genetic Data to Injury Risk Assessment in Athletes," *PLOS One* 10 (2014): e0122676.

12. L. M. Guth and S. M. Roth, "Genetic Influence on Athletic Performance," *Current Opinion in Pediatrics* 25 (2013): 653–658.

13. M. Gumpenberger et al., "Remodeling the Skeletal Muscle Extracellular Matrix in Older Age-Effects of Acute Exercise Stimuli on Gene Expression," *International Journal of Molecular Sciences* 21 (2020).

14. R. Cacabelos, N. Cacabelos, and J. C. Carril, "The Role of Pharmacogenomics in Adverse Drug Reactions," *Expert Review of Clinical Pharmacology* 12 (2019): 407–442; M. V. Relling et al., "The Clinical Pharmacogenetics Implementation Consortium: 10 Years Later," *Clinical Pharmacology & Therapeutics* 107 (2020): 171–175.

15. W. T. Nicholson et al., "Considerations When Applying Pharmacogenomics to Your Practice," *Mayo Clinic Proceedings* 96 (2021): 218–230; K. Krebs and L. Milani, "Translating Pharmacogenomics into Clinical Decisions: Do Not Let the Perfect Be the Enemy of the Good," *Human Genomics* 13 (2019): 39.

16. J. Zaritsky, dir., *Do You Really Want to Know?* Optic Nerve Films, 2012, https://www.doyoureallywanttoknowfilm.com/subjects/.

17. M. Gumpenberger et al., "Remodeling the Skeletal Muscle Extracellular Matrix in Older Age-Effects of Acute Exercise Stimuli on Gene Expression," *International Journal of Molecular Sciences* 21 (2020): 7089.

18. W. Kopp, "How Western Diet and Lifestyle Drive the Pandemic of Obesity and Civilization Diseases." *Diabetes, Metabolic Syndrome, and Obesity* 12 (2019): 2221–2236.

19. M. W. Gillman, "Predicting Prediabetes and Diabetes: Can We Do It? Is It Worth It?" *Archives of Pediatrics & Adolescent Medicine* 164 (2010): 198–199.

20. D. Shin, K. Kongpakpaisarn, and C. Bohra, "Trends in the Prevalence of Metabolic Syndrome and Its Components in the United States, 2007–2014, *International Journal of Cardiology* 259 (2018): 216–219.

21. J. D. Cohen et al., "Detection and Localization of Surgically Resectable Cancers with a Multi-Analyte Blood Test," *Science* 359 (2018): 926–930.

22. A. T. Magis et al., "Untargeted Longitudinal Analysis of a Wellness Cohort Identifies Markers of Metastatic Cancer Years Prior to Diagnosis," *Scientific Reports* 10 (2020): 16275.

23. M. E. Levine, "Modeling the Rate of Senescence: Can Estimated Biological Age Predict Mortality More Accurately Than Chronological Age?"

The Journals of Gerontology Series A: Biological Sciences & Medical Sciences 68 (2013): 667–674.

24. J. C. Earls et al., "Multi-Omic Biological Age Estimation and Its Correlation with Wellness and Disease Phenotypes: A Longitudinal Study of 3,558 Individuals," *The Journals of Gerontology Series A: Biological Sciences & Medical Sciences* 74 (2019): S52–S60.

25. E. Lim, J. Miyamura, and J. J. Chen, "Racial/Ethnic-Specific Reference Intervals for Common Laboratory Tests: A Comparison among Asians, Blacks, Hispanics, and White," *Hawai'i Journal of Medicine & Public Health* 74 (2015): 302–310.

26. N. Zubair et al., "Genetic Predisposition Impacts Clinical Changes in a Lifestyle Coaching Program," *Scientific Reports* 9 (2019): 6805.

27. M. Wainberg et al., "Multiomic Blood Correlates of Genetic Risk Identify Presymptomatic Disease Alterations," *Proceedings of the National Academy of Sciences of the United States of America* 117 (2020): 21813–21820.

28. J. Riancho et al., "The Increasing Importance of Environmental Conditions in Amyotrophic Lateral Sclerosis," *International Journal of Biometeorology* 62 (2018): 1361–1374.

29. A. M. Valdes et al., "Role of the Gut Microbiota in Nutrition and Health," *The BMJ* 361 (2018): k2179.

30. C. S. Liou et al., "A Metabolic Pathway for Activation of Dietary Glucosinolates by a Human Gut Symbiont," *Cell* 180 (2020): 717–728.e719.

31. B. Javdan et al., "Personalized Mapping of Drug Metabolism by the Human Gut Microbiome," *Cell* 181 (2020): 1661–1679.e1622.

32. T. Wilmanski et al., "Gut Microbiome Pattern Reflects Healthy Ageing and Predicts Survival in Humans," *Nature Metabolism* 3 (2021): 274–286.

33. Y. He et al., "Regional Variation Limits Applications of Healthy Gut Microbiome Reference Ranges and Disease Models," *Nature Medicine* 24 (2018): 1532–1535.

34. T. Wilmanski et al., "Blood Metabolome Predicts Gut Microbiome Alpha-Diversity in Humans," *Nature Biotechnology* 37 (2019): 1217–1228.

35. B. Reeder and A. David, "Health at Hand: A Systematic Review of Smart Watch Uses for Health and Wellness," *Journal of Biomedical Informatics* 63 (2016): 269–276.

36. E. O'Brien, C. Hart, and R. R. Wing, "Discrepancies between Self-Reported Usual Sleep Duration and Objective Measures of Total Sleep Time in Treatment-Seeking Overweight and Obese Individuals," *Behavioral Sleep Medicine* 14 (2016): 539–549; M. R. Janevic, S. J. McLaughlin, and C. M. Connell, "Overestimation of Physical Activity among a Nationally Representative Sample of Underactive Individuals with Diabetes," *Medical Care* 50 (2012): 441–445.

37. M. Abdelsayed, M. Ruprai, and P. C. Ruben, "The Efficacy of Ranolazine on E1784k Is Altered by Temperature and Calcium," *Scientific Reports* 8 (2018): 3643.

38. M. V. Perez et al., "Large-Scale Assessment of a Smartwatch to Identify Atrial Fibrillation," *New England Journal of Medicine* 381 (2019): 1909–1917.

39. Human Phenome Institute, "The 2nd International Symposium of Human Phenomics Was Held in Shanghai," November 11, 2018, Fudan University, https://hupi.fudan.edu.cn/en/content.jsp?urltype=news.NewsContentUrl&wbtreeid=1041&wbnewsid=1065.

4. Measuring and Tracking Health

1. G. Z. Segal, *Getting There: A Book of Mentors* (New York: Harry N. Abrams, 2015).

2. N. Bomey, "Old Cars Everywhere: Average Vehicle Age Hits All-Time High," *USA Today*, June 28, 2019.

3. K. Jeyaraman, "Diabetic-Foot Complications in American and Australian Continents," in *Diabetic Foot Ulcer: An Update*, ed. M. Zubair et al. (Springer Singapore, 2021), 41–59.

4. Centers for Disease Control and Prevention, *Chronic Kidney Disease in the United States, 2021* (US Department of Health and Human Services & Centers for Disease Control and Prevention, 2021).

5. Centers for Disease Control and Prevention, "Awareness of Prediabetes—United States, 2005–2010," *The Morbidity and Mortality Weekly Report* 62 (2013): 209–212.

6. W. Bracamonte-Baran and D. Cihakova, "Cardiac Autoimmunity: Myocarditis," *Advances in Experimental Medicine and Biology* 1003 (2017): 187–221.

7. Y. Zheng et al., "Self-Weighing in Weight Management: A Systematic Literature Review." *Obesity* 23 (2015): 256–265.

8. U. A. R. Chaudhry et al., "The Effects of Step-Count Monitoring Interventions on Physical Activity: Systematic Review and Meta-Analysis of Community-Based Randomised Controlled Trials in Adults," *International Journal of Behavioral Nutrition and Physical Activity* 17 (2020): 129.

9. P. J. Taylor et al., "Efficacy of Real-Time Continuous Glucose Monitoring to Improve Effects of a Prescriptive Lifestyle Intervention in Type 2 Diabetes: A Pilot Study," *Diabetes Therapy* 10 (2019): 509–522.

10. D. Zeevi et al., "Personalized Nutrition by Prediction of Glycemic Responses," *Cell* 163 (2015): 1079–1094.

11. O. Liebermann, M. Schwartz, and A. Tal, "Israel Is Deploying Spy Technology to Track the Virus, Prompting Fears of Privacy Invasion," CNN.com, March 18, 2020, https://www.cnn.com/2020/03/18/tech/israel-coronavirus-technology-intl/index.html.

12. M. Ilyushina, "How Russia Is Using Authoritarian Tech to Curb Coronavirus," CNN.com, March 29, 2020, https://www.cnn.com/2020/03/29/europe/russia-coronavirus-authoritarian-tech-intl/index.html.

13. "France Launches Vaccine Pass for Cultural Venues," France 24, July 21, 2021, https://www.france24.com/en/live-news/20210721-france-launches-vaccine-pass-for-cultural-venues-1.

14. B. Fung, "Trump Administration Wants to Use Americans' Location Data to Track the Coronavirus," CNN Business, March 18, 2020, https://www.cnn.com/2020/03/18/tech/us-government-location-data-coronavirus/index.html.

15. J. Liu and H. Zhao "Privacy Lost: Appropriating Surveillance Technology in China's Fight against COVID-19," *Business Horizons* 64 (2021): 743–756.

16. D. Levin et al., "Diversity and Functional Landscapes in the Microbiota of Animals in the Wild," *Science* 372 (2021).

17. B. Auxier et al., "Americans and Privacy: Concerned, Confused and Feeling Lack of Control over Their Personal Information," Pew Research Center, November 15, 2019, https://www.pewresearch.org/internet/2019/11/15/americans-and-privacy-concerned-confused-and-feeling-lack-of-control-over-their-personal-information/.

18. M. DoBias, "Ron Paul's Lonely Opposition," *Modern Healthcare*, May 12, 2008.

19. Y. Joly et al., "Looking Beyond GINA: Policy Approaches to Address Genetic Discrimination," *Annual Review of Genomics and Human Genetics* 21 (2020): 491–507.

20. N. D. Price et al., "Highly Accurate Two-Gene Classifier for Differentiating Gastrointestinal Stromal Tumors and Leiomyosarcomas," *Proceedings of the National Academy of Sciences of the United States of America* 104 (2007): 3414–3419.

21. N. D. Price et al., "Highly Accurate Two-Gene Classifier for Differentiating Gastrointestinal Stromal Tumors and Leiomyosarcomas," 3414–3419.

22. H. O. Adami, "Time to Abandon Early Detection Cancer Screening," *European Journal of Clinical Investigation* 49 (2019): e13062.

23. G. B. Taksler, N. L. Keating, and M. B. Rothberg, "Implications of False-Positive Results for Future Cancer Screenings," *Cancer* 124 (2018): 2390–2398.

24. Taksler, Keating, and Rothberg, "Implications of False-Positive Results."

25. F. Perraudeau et al., "Improvements to Postprandial Glucose Control in Subjects with Type 2 Diabetes: A Multicenter, Double Blind, Randomized Placebo-Controlled Trial of a Novel Probiotic Formulation," *BMJ Open Diabetes Research & Care* 8 (2020).

26. V. R. Aroda et al., "Metformin for Diabetes Prevention: Insights Gained from the Diabetes Prevention Program / Diabetes Prevention Program Outcomes Study," *Diabetologia* 60 (2017): 1601–1611; H. Wang et al., "Metformin and Berberine, Two Versatile Drugs in Treatment of Common Metabolic Diseases," *Oncotarget* 9 (2018): 10135–10146.

27. J. S. Roberts, "Assessing the Psychological Impact of Genetic Susceptibility Testing," *Hastings Center Report* 49, suppl. 1 (2019): S38–S43.

28. K. D. Christensen et al., "Behavioral and Psychological Impact of Genome Sequencing: A Pilot Randomized Trial of Primary Care and Cardiology Patients," *NPJ Genomic Medicine* 6 (2021): 72.

29. R. C. Green et al., "Disclosure of APOE Genotype for Risk of Alzheimer's Disease," *New England Journal of Medicine* 361 (2009): 245–254.

30. H. Hua et al., "A Wipe-Based Stool Collection and Preservation Kit for Microbiome Community Profiling," *Frontiers in Immunology* 13 (2022).

31. R. Nunn, J. Parsons, and J. Shambaugh, *A Dozen Facts about the Economics of the US Health-Care System*, The Brookings Institution, March 10, 2020, https://www.brookings.edu/research/a-dozen-facts-about-the-economics-of-the-u-s-health-care-system/.

32. L. W. Sullivan and I. Suez Mittman, "The State of Diversity in the Health Professions a Century after Flexner," *Academic Medicine* 85 (2010): 246–253.

33. B. Herman, "Health Care CEOs Took Home $2.6 Billion in 2018," Axios, May 16, 2019, https://www.axios.com/2019/05/16/health-care-ceo-pay-compensation-stock-2018.

34. M. Sullivan, "Federal Budget Grants $1.8 Billion to Alzheimer's and Dementia Research," *Clinical Psychiatry News*, March 30, 2018.

5. A New Way to Think about How Old We Are

1. D. Santesmasses et al., "COVID-19 Is an Emergent Disease of Aging," *Aging Cell* 19 (2020): e13230.

2. I. Seim et al., "The Transcriptome of the Bowhead Whale *Balaena mysticetus* Reveals Adaptations of the Longest-Lived Mammal," *Aging* 6 (2014): 879–899.

3. J. Nielsen et al., "Eye Lens Radiocarbon Reveals Centuries of Longevity in the Greenland Shark (*Somniosus microcephalus*)," *Science* 353 (2016): 702–704.

4. J. P. Mackenbach et al., "Gains in Life Expectancy after Elimination of Major Causes of Death: Revised Estimates Taking into Account the Effect of Competing Causes," *Journal of Epidemiology and Community Health* 53 (1999): 32–37.

5. Z. He et al., "Prevalence of Multiple Chronic Conditions among Older Adults in Florida and the United States: Comparative Analysis of the OneFlorida Data Trust and National Inpatient Sample," *Journal of Medical Internet Research* 20 (2018): e137.

6. P. Yang, "What Happens When We All Live to 100?" *The Atlantic*, October 2014.

7. S. M. Burstein and C. E. Finch, "Longevity Examined: An Ancient Greek's Very Modern Views on Ageing," *Nature* 560 (2018): 430.

8. M. E. Matzkin et al., "Hallmarks of Testicular Aging: The Challenge of Anti-Inflammatory and Antioxidant Therapies Using Natural and/or Pharmacological Compounds to Improve the Physiopathological Status of the Aged Male Gonad," *Cells* 10 (2021); M. Lemoine, "The Evolution of the Hallmarks of Aging," *Frontiers in Genetics* 12 (2021): 693071; G. R. Guimaraes et al., "Hallmarks of Aging in Macrophages: Consequences to Skin Inflammaging," *Cells* 10 (2021); M. Mittelbrunn and G. Kroemer, "Hallmarks of T Cell Aging," *Nature Immunology* 22 (2021): 687–698; S. van der Rijt et al., "Integrating the Hallmarks of Aging throughout the Tree of Life: A Focus on Mitochondrial Dysfunction," *Frontiers in Cell and Developmental Biology* 8 (2020): 594416; F. Guerville et al., "Revisiting the Hallmarks of Aging to Identify Markers of Biological Age," *Journal of Prevention of Alzheimer's Disease* 7 (2020): 56–64; S. Kaushik et al., "Autophagy and the Hallmarks of Aging," *Ageing Research Reviews* 72 (2021): 101468; L. B. Boyette and R. S. Tuan, "Adult Stem Cells and Diseases of Aging," *Journal of Clinical Medicine* 3 (2014): 88–134; L. P. Rodrigues et al., "Hallmarks of Aging and Immunosenescence: Connecting the Dots," *Cytokine Growth Factor Reviews* 59 (2021): 9–21; S. Dodig, I. Cepelak, and I. Pavic, "Hallmarks of Senescence and Aging," *Biochemia Medica* 29 (2019): 030501.

9. C. Lopez-Otin, "The Hallmarks of Aging," *Cell* 153 (2013): 1194–1217.

10. Our interest in Sinclair's work, documented in his thought-provoking book *Lifespan*, preceded our decision to work with his co-writer, journalist Matthew D. LaPlante, but it is appropriate to acknowledge this conflict.

11. A. Martin-Montalvo et al., "Metformin Improves Healthspan and Lifespan in Mice," *Nature Communications* 4 (2013): 2192.

12. P. Klemera and S. Doubal, "A New Approach to the Concept and Computation of Biological Age," *Mechanisms of Ageing and Development* 127 (2006): 240–248.

13. J. C. Earls, "Multi-Omic Biological Age Estimation and Its Correlation with Wellness and Disease Phenotypes: A Longitudinal Study of 3,558 Individuals," *The Journals of Gerontology Series A: Biological Sciences & Medical Sciences* 74 (2019): S52–S60.

14. C. J. Caspersen et al., "Aging, Diabetes, and the Public Health System in the United States," *American Journal of Public Health* 102 (2012): 1482–1497.

15. Y. Lu et al., "Reprogramming to Recover Youthful Epigenetic Information and Restore Vision," *Nature* 588 (2020): 124–129; S. A. Villeda et al., "Young Blood Reverses Age-Related Impairments in Cognitive Function and Synaptic Plasticity in Mice," *Nature Medicine* 20 (2014): 659–663; J. P. de Magalhaes and A. Ocampo, "Cellular Reprogramming and the Rise of Rejuvenation Biotech," *Trends in Biotechnology* (2022).

16. Z. Liu et al., "A New Aging Measure Captures Morbidity and Mortality Risk across Diverse Subpopulations from NHANES IV: A Cohort Study," *PLOS Medicine* 15 (2018): e1002718; P. J. Brown et al., "Biological Age, Not Chronological Age, Is Associated with Late-Life Depression," *The Journals of Gerontology Series A: Biological Sciences & Medical Sciences* 73 (2018): 1370–1376; M. E. Levine, "Modeling the Rate of Senescence: Can Estimated Biological Age Predict Mortality More Accurately Than Chronological Age?" *The Journals of Gerontology Series A: Biological Sciences & Medical Sciences* 68 (2013): 667–674.

17. Liu et al., "A New Aging Measure."

18. J. J. Arnett, "Emerging Adulthood. A Theory of Development from the Late Teens through the Twenties," *American Psychologist* 55 (2000): 469–480.

19. S. Horvath and K. Raj, "DNA Methylation-Based Biomarkers and the Epigenetic Clock Theory of Ageing," *Nature Reviews Genetics* 19 (2018): 371–384.

20. A. Quach et al., "Epigenetic Clock Analysis of Diet, Exercise, Education, and Lifestyle Factors," *Aging* 9 (2017): 419–446.

21. P. D. Fransquet et al., "The Epigenetic Clock as a Predictor of Disease and Mortality Risk: A Systematic Review and Meta-Analysis," *Clinical Epigenetics* 11 (2019): 62.

22. A. Harris, "How Old Are You Really? Elysium Health Will Tell You—For $500," Fast Company, November 4, 2019, https://www.fastcompany.com/90406604/how-old-are-you-really-elysium-health-will-tell-you-for-500.

23. Liu et al., "A New Aging Measure."

24. J. Yoshino, J. A. Baur, and S.-I. Imai, "NAD^+ Intermediates: The Biology and Therapeutic Potential of NMN and NR," *Cell Metabolism* 27 (2018): 513–528.

25. J. Ratajczak et al., "NRK1 Controls Nicotinamide Mononucleotide and Nicotinamide Riboside Metabolism in Mammalian Cells," *Nature Communications* 7 (2016): 13103.

26. S. A. Trammell et al., "Nicotinamide Riboside Is Uniquely and Orally Bioavailable in Mice and Humans," *Nature Communications* 7 (2016): 12948.

27. O. K. Reiten et al., "Preclinical and Clinical Evidence of NAD^+ Precursors in Health, Disease, and Ageing," *Mechanisms of Ageing and Development* 199 (2021): 111567.

28. B. E. Kang et al., "Implications of NAD^+ Boosters in Translational Medicine," *European Journal of Clinical Investigation* 50 (2020): e13334.

29. M. V. Blagosklonny, "Rapamycin for Longevity: Opinion Article," *Aging* 11 (2019): 8048–8067.

30. D. E. Harrison et al., "Rapamycin Fed Late in Life Extends Lifespan in Genetically Heterogeneous Mice," *Nature* 460 (2009): 392–395.

31. Blagosklonny, "Rapamycin for Longevity."

32. J. Y. An et al., "Rapamycin Rejuvenates Oral Health in Aging Mice," *eLife* 9 (2020).

33. S. R. Urfer et al., "A Randomized Controlled Trial to Establish Effects of Short-Term Rapamycin Treatment in 24 Middle-Aged Companion Dogs," *Geroscience* 39 (2017): 117–127.

34. G. M. Fahy et al., "Reversal of Epigenetic Aging and Immunosenescent Trends in Humans," *Aging Cell* 18 (2019): e13028.

35. M. L. Vance, "Can Growth Hormone Prevent Aging?" *New England Journal of Medicine* 348 (2003): 779–780.

6. Keeping Our Minds Healthy for Life

1. R. Stepler, "World's Centenarian Population Projected to Grow Eightfold by 2050." Pew Research Center, April 21, 2016, https://www.pewresearch.org/fact-tank/2016/04/21/worlds-centenarian-population-projected-to-grow-eightfold-by-2050/.

2. O. K. Reiten et al., "Preclinical and Clinical Evidence of NAD^+ Precursors in Health, Disease, and Ageing," *Mechanisms of Ageing and Development* 199 (2021): 111567.

3. G. Kempermann and F. H. Gage, "New Nerve Cells for the Adult Brain," *Scientific American* 280 (1999): 48–53.

4. H. W. Mahncke, A. Bronstone, and M. M. Merzenich, "Brain Plasticity and Functional Losses in the Aged: Scientific Bases for a Novel Intervention," *Progress in Brain Research* 157 (2006): 81–109.

5. Mahncke, Bronstone, and Merzenich, "Brain Plasticity and Functional Losses in the Aged."

6. M. Raab, J. Johnson, and H. Heekeren, *Mind and Motion: The Bidirectional Link between Thought and Action* (Elsevier Science, 2009).

7. D. A. Lombardi, W. J. Horrey, and T. K. Courtney, "Age-Related Differences in Fatal Intersection Crashes in the United States," *Accident Analysis and Prevention* 99 (2017): 20–29.

8. C. Maynard et al., "Disability Rating, Age at Death, and Cause of Death in U.S. Veterans with Service-Connected Conditions," *Military Medicine* 183 (2018): e371–e376.

9. M. M. Merzenich, T. M. Van Vleet, and M. Nahum, "Brain Plasticity-Based Therapeutics," *Frontiers in Human Neuroscience* 8 (2014): 385.

10. M. Merzenich and K. Ball, "At the Cusp of Solving Cognitive Aging?" Medium, March 15, 2017, https://medium.com/@MichaelMerzenich/at-the-cusp-of-solving-cognitive-aging-9907a8b7775f.

11. H. W. Mahncke et al., "Memory Enhancement in Healthy Older Adults Using a Brain Plasticity-Based Training Program: A Randomized, Controlled Study," *Proceedings of the National Academy of Sciences of the United States of America* 103 (2006): 12523–12528.

12. H. K. Lee et al., "Home-Based, Adaptive Cognitive Training for Cognitively Normal Older Adults: Initial Efficacy Trial," *The Journals of Gerontology Series B: Psychological Sciences and Social Sciences* 75 (2020): 1144–1154.

13. Lee et al., "Home-Based, Adaptive Cognitive Training."

14. C. Hardcastle et al., "Higher-Order Resting State Network Association with the Useful Field of View Task in Older Adults," *Geroscience* 44 (2022): 131–145.

15. D. J. Simons et al., "Do 'Brain-Training' Programs Work?" *Psychological Science in the Public Interest* 17 (2016): 103–186.

16. J. Mishra et al., "Training Sensory Signal-to-Noise Resolution in Children with ADHD in a Global Mental Health Setting," *Translational Psychiatry* 6 (2016): e781; E. M. Boutzoukas et al., "Higher White Matter Hyperintensity Load Adversely Affects Pre-Post Proximal Cognitive Training Performance in Healthy Older Adults," *Geroscience* 44 (2022): 1441–1455; C. Hardcastle et al., "Proximal Improvement and Higher-Order Resting State Network Change after Multidomain Cognitive Training Intervention in Healthy Older Adults," *Geroscience* 44 (2022): 1011–1027; J. D. Edwards et al., "Speed of Processing Training Results in Lower Risk of Dementia," *Alzheimer's & Dementia* 3 (2017): 603–611; R. Manenti et al., "Transcranial Direct Current Stimulation Combined with Cognitive Training for the Treatment of Parkinson Disease: A Randomized, Placebo-Controlled Study," *Brain Stimulation* 11 (2018): 1251–1262; J. D. Edwards et al., "Randomized Trial of Cognitive Speed of Processing Training in Parkinson Disease," *Neurology* 81 (2013): 1284–1290.

17. List kept current at brainhq.com (https://www.brainhq.com/world-class-science/information-researchers/).

18. J. Zimman, "Tom Brady's (No Longer) Secret Weapon: BrainHQ," brainhq.com, January 29, 2017, https://www.brainhq.com/blog/tom-bradys-no-longer-secret-weapon-brainhq/.

19. S. L. Miller et al., "An Investigation of Computer-Based Brain Training on the Cognitive and EEG Performance of Employees," *2019 Annual International Conference of the IEEE Engineering in Medicine and Biology Society (EMBC)* 2019 (2019): 518–521.

20. J. Hamilton et al., "Can Cognitive Training Improve Shoot/Don't-Shoot Performance? Evidence from Live Fire Exercises," *American Journal of Psychiatry* 132 (2019): 179–194.

21. J. Walters et al., "Improving Attentiveness: Effect of Cognitive Training on Sustained Attention Measures," *Professional Safety* 64, no. 4 (2019): 31–35.

22. T. M. Shah et al., "Enhancing Cognitive Functioning in Healthy Older Adults: A Systematic Review of the Clinical Significance of Commercially Available Computerized Cognitive Training in Preventing Cognitive Decline," *Neuropsychology Review* 27 (2017): 62–80.

23. K. Rehfeld et al., "Dancing or Fitness Sport? The Effects of Two Training Programs on Hippocampal Plasticity and Balance Abilities in Healthy Seniors," *Frontiers in Human Neuroscience* 11 (2017): 305.

24. G. Bubbico et al., "Effects of Second Language Learning on the Plastic Aging Brain: Functional Connectivity, Cognitive Decline, and Reorganization," *Frontiers in Neuroscience* 13 (2019): 423.

25. G. M. Bidelman and C. Alain, "Musical Training Orchestrates Coordinated Neuroplasticity in Auditory Brainstem and Cortex to Counteract Age-Related Declines in Categorical Vowel Perception," *Journal of Neuroscience* 35 (2015): 1240–1249; C. E. James et al., "Train the Brain with Music (TBM): Brain Plasticity and Cognitive Benefits Induced by Musical Training in Elderly People in Germany and Switzerland, a Study Protocol for an RCT Comparing Musical Instrumental Practice to Sensitization to Music," *BMC Geriatrics* 20 (2020): 418.

26. S. Mondello et al., "Blood-Based Diagnostics of Traumatic Brain Injuries," *Expert Review of Molecular Diagnostics* 11 (2011): 65–78; N. Aghakhani, "Relationship between Mild Traumatic Brain Injury and the Gut Microbiome: A Scoping Review," *Journal of Neuroscience Research* 100 (2022): 827–834; C. S. Zhu et al., "A Review of Traumatic Brain Injury and the Gut Microbiome: Insights into Novel Mechanisms of Secondary Brain Injury and Promising Targets for Neuroprotection," *Brain Sciences* 8 (2018).

27. B. D. Needham, R. Kaddurah-Daouk, and S. K. Mazmanian, "Gut Microbial Molecules in Behavioural and Neurodegenerative Conditions," *Na-*

ture Reviews Neuroscience 21 (2020): 717–731; L. H. Morais, H. L. Schreiber IV, and S. K. Mazmanian, "The Gut Microbiota–Brain Axis in Behaviour and Brain Disorders," *Nature Reviews Microbiology* 19 (2021): 241–255.

28. P. H. Croll et al., "Better Diet Quality Relates to Larger Brain Tissue Volumes: The Rotterdam Study," *Neurology* 90 (2018): e2166–e2173.

29. B. Sharma, D. W. Lawrence, and M. G. Hutchison, "Branched Chain Amino Acids (BCAAs) and Traumatic Brain Injury: A Systematic Review," *The Journal of Head Trauma Rehabilitation* 33 (2018): 33–45; A. Wu, Z. Ying, and F. Gomez-Pinilla, "Dietary Curcumin Counteracts the Outcome of Traumatic Brain Injury on Oxidative Stress, Synaptic Plasticity, and Cognition," *Experimental Neurology* 197 (2006): 309–317; J. E. Bailes and V. Patel, "The Potential for DHA to Mitigate Mild Traumatic Brain Injury," *Military Medicine* 179 (2014): 112–116; R. Vink et al., "Magnesium Attenuates Persistent Functional Deficits Following Diffuse Traumatic Brain Injury in Rats," *Neuroscience Letters* 336 (2003): 41–44; H. Yang et al., "Ketone Bodies in Neurological Diseases: Focus on Neuroprotection and Underlying Mechanisms," *Frontiers in Neurology* 10 (2019): 585; M. E. Hoffer et al., "Amelioration of Acute Sequelae of Blast Induced Mild Traumatic Brain Injury by N-Acetyl Cysteine: A Double-Blind, Placebo Controlled Study," *PLOS One* 8 (2013): e54163; U. Sonmez et al., "Neuroprotective Effects of Resveratrol against Traumatic Brain Injury in Immature Rats," *Neuroscience Letters* 420 (2007): 133–137; M. R. Hoane, J. G. Wolyniak, and S. L. Akstulewicz, "Administration of Riboflavin Improves Behavioral Outcome and Reduces Edema Formation and Glial Fibrillary Acidic Protein Expression after Traumatic Brain Injury," *Journal of Neurotrauma* 22 (2005): 1112–1122; T. L. Roth et al., "Transcranial Amelioration of Inflammation and Cell Death after Brain Injury," *Nature* 505 (2014): 223–228.

30. L. Mosconi, *Brain Food: The Surprising Science of Eating for Cognitive Power* (Avery, 2018).

31. M. P. T. Ylilauri et al., "Associations of Dietary Choline Intake with Risk of Incident Dementia and with Cognitive Performance: The Kuopio Ischaemic Heart Disease Risk Factor Study," *American Journal of Clinical Nutrition* 110 (2019): 1416–1423.

32. C. W. Cotman, N. C. Berchtold, and L. A. Christie, "Exercise Builds Brain Health: Key Roles of Growth Factor Cascades and Inflammation," *Trends in Neurosciences* 30 (2007): 464–472.

33. Z. Wang et al., "Poor Sleep Quality Is Negatively Associated with Low Cognitive Performance in General Population Independent of Self-Reported Sleep Disordered Breathing," *BMC Public Health* 22 (2022): 3; K. R. Johannsdottir et al., "Objective Measures of Cognitive Performance in Sleep Disorder Research," *Sleep Medicine Clinics* 16 (2021): 575–593; T. Csipo et al., "Sleep

Deprivation Impairs Cognitive Performance, Alters Task-Associated Cerebral Blood Flow and Decreases Cortical Neurovascular Coupling-Related Hemodynamic Responses," *Scientific Reports* 11 (2021): 20994.

34. J. A. Mortimer et al., "Changes in Brain Volume and Cognition in a Randomized Trial of Exercise and Social Interaction in a Community-Based Sample of Non-Demented Chinese Elders," *Journal of Alzheimer's Disease* 30 (2012): 757–766.

35. E. R. de Kloet et al., "Stress and Depression: A Crucial Role of the Mineralocorticoid Receptor," *Journal of Neuroendocrinology* 28 (2016).

7. The Long Goodbye

1. C. Dwyer, "Pfizer Halts Research into Alzheimer's and Parkinson's Treatments," NPR, January 8, 2018, https://www.npr.org/sections/thetwo-way/2018/01/08/576443442/pfizer-halts-research-efforts-into-alzheimers-and-parkinsons-treatments.

2. S. Makin, "The Amyloid Hypothesis on Trial," *Nature* 559 (2018): S4–S7.

3. C. Kichenbrand et al., "Brain Abscesses and Intracranial Empyema Due to Dental Pathogens: Case Series," *International Journal of Surgery Case Reports* 69 (2020): 35–38.

4. J. Hellmuth, "Can We Trust *The End of Alzheimer's?*" *Lancet Neurology* 19, no. 5 (2020): 389–390.

5. D. E. Bredesen et al., "Reversal of Cognitive Decline in Alzheimer's Disease," *Aging* 8 (2016): 1250–1258.

6. K. Toups et al., "Precision Medicine Approach to Alzheimer's Disease: Successful Pilot Project," *Journal of Alzheimer's Disease* 88 (2022): 1411–1421.

7. R. V. Rao et al., "RECODE: A Personalized, Targeted, Multi-Factorial Therapeutic Program for Reversal of Cognitive Decline," *Biomedicines* 9 (2021).

8. N. Coley et al., "Adherence to Multidomain Interventions for Dementia Prevention: Data from the FINGER and MAPT Trials," *Alzheimer's & Dementia* 15 (2019): 729–741; A. Rosenberg et al., "Multidomain Lifestyle Intervention Benefits a Large Elderly Population at Risk for Cognitive Decline and Dementia Regardless of Baseline Characteristics: The FINGER Trial," *Alzheimer's & Dementia* 14 (2018): 263–270; T. Ngandu et al., "A 2 Year Multidomain Intervention of Diet, Exercise, Cognitive Training, and Vascular Risk Monitoring versus Control to Prevent Cognitive Decline in At-Risk Elderly People (FINGER): A Randomised Controlled Trial," *Lancet* 385 (2015): 2255–2263.

9. T. Ngandu, interview at the 31st International Conference of Alzheimer's Disease International, Budapest, Hungary, April 22, 2016, *VJ Dementia: The Video Journal of Dementia.*

8. Deciphering Dementia

1. S. Alig et al., "Impact of Age on Genetics and Treatment Efficacy in Follicular Lymphoma," *Haematologica* 103 (2018): e364–e367.
2. Alig et al., "Impact of Age on Genetics and Treatment Efficacy."
3. C. W. Huang et al., "Cerebral Perfusion Insufficiency and Relationships with Cognitive Deficits in Alzheimer's Disease: A Multiparametric Neuroimaging Study," *Scientific Reports* 8 (2018): 1541.
4. G. A. Edwards III et al., "Modifiable Risk Factors for Alzheimer's Disease," *Frontiers in Aging Neuroscience* 11 (2019): 146.
5. G. Livingston et al., "Dementia Prevention, Intervention, and Care: 2020 Report of the Lancet Commission," *Lancet* 396 (2020): 413–446.
6. Q. Meng, M. S. Lin, and I. S. Tzeng, "Relationship between Exercise and Alzheimer's Disease: A Narrative Literature Review," *Frontiers in Neuroscience* 14 (2020): 131.
7. N. Coley et al., "Adherence to Multidomain Interventions for Dementia Prevention: Data from the FINGER and MAPT Trials," *Alzheimer's & Dementia* 15 (2019): 729–741; A. Rosenberg et al., "Multidomain Lifestyle Intervention Benefits a Large Elderly Population at Risk for Cognitive Decline and Dementia Regardless of Baseline Characteristics: The FINGER Trial," *Alzheimer's & Dementia* 14 (2018): 263–270; T. Ngandu et al., "A 2 Year Multidomain Intervention of Diet, Exercise, Cognitive Training, and Vascular Risk Monitoring versus Control to Prevent Cognitive Decline in At-Risk Elderly People (FINGER): A Randomised Controlled Trial," *Lancet* 385 (2015): 2255–2263.
8. J. Fang et al., "Endophenotype-Based in Silico Network Medicine Discovery Combined with Insurance Record Data Mining Identifies Sildenafil as a Candidate Drug for Alzheimer's Disease," *Nature Aging* 1 (2021): 1175–1188.
9. S. Seshadri, D. A. Drachman, and C. F. Lippa, "Apolipoprotein E Epsilon 4 Allele and the Lifetime Risk of Alzheimer's Disease: What Physicians Know, and What They Should Know," *Archives Neurology* 52 (1995): 1074–1079; M. E. Belloy, V. Napolioni, and M. D. Greicius, "A Quarter Century of APOE and Alzheimer's Disease: Progress to Date and the Path Forward," *Neuron* 101 (2019): 820–838.
10. Z. Li et al., "APOE2: Protective Mechanism and Therapeutic Implications for Alzheimer's Disease," *Molecular Neurodegeneration* 15 (2020): 63; S. Suri et al., "The Forgotten APOE Allele: A Review of the Evidence and

Suggested Mechanisms for the Protective Effect of APOE Varepsilon2," *Neuroscience & Biobehavioral Reviews* 37 (2013): 2878–2886.

11. R. W. Mahley, "Central Nervous System Lipoproteins: APOE and Regulation of Cholesterol Metabolism," *Arteriosclerosis, Thrombosis, and Vascular Biology* 36 (2016): 1305–1315.

12. A. R. Garcia et al., "APOE4 Is Associated with Elevated Blood Lipids and Lower Levels of Innate Immune Biomarkers in a Tropical Amerindian Subsistence Population," *eLife* 10 (2021); Y. Huang et al., "Apolipoprotein E2 Reduces the Low-Density Lipoprotein Level in Transgenic Mice by Impairing Lipoprotein Lipase-Mediated Lipolysis of Triglyceride-Rich Lipoproteins," *Journal of Biological Chemistry* 273 (1998): 17483–17490.

13. B. Das and R. Yan, "A Close Look at BACE1 Inhibitors for Alzheimer's Disease Treatment," *CNS Drugs* 33 (2019): 251–263.

14. M. R. Egan et al., "Randomized Trial of Verubecestat for Mild-to-Moderate Alzheimer's Disease," *New England Journal of Medicine* 378 (2018): 1691–1703.

15. N. M. Moussa-Pacha et al., "BACE1 Inhibitors: Current Status and Future Directions in Treating Alzheimer's Disease," *Medicinal Research Reviews* 40 (2020): 339–384.

16. P. Belluck, "F.D.A. Panel Declines to Endorse Controversial Alzheimer's Drug," *New York Times*, November 6, 2020, https://www.nytimes.com/2020/11/06/health/aducanumab-alzheimers-drug-fda-panel.html.

17. This quote is frequently attributed to Mark Twain. As is often the case, there is no record of him having written or said these words.

18. M. S. Parihar and G. J. Brewer, "Amyloid-Beta as a Modulator of Synaptic Plasticity," *Journal of Alzheimer's Disease* 22 (2010): 741–763.

19. A. A. Apostolopoulou and A. C. Lin, "Mechanisms Underlying Homeostatic Plasticity in the Drosophila Mushroom Body in Vivo," *Proceedings of the National Academy of Sciences of the United States of America* 117 (2020): 16606–16615.

20. M. Oka et al., "Ca^{2+}/Calmodulin-Dependent Protein Kinasef II Promotes Neurodegeneration Caused by Tau Phosphorylated at Ser262/356 in a Transgenic *Drosophila* Model of Tauopathy," *Journal of Biochemistry* 162 (2017): 335–342.

21. A. R. Koudinov and N. V. Koudinova, "Cholesterol Homeostasis Failure as a Unifying Cause of Synaptic Degeneration," *Journal of the Neurological Sciences* 229–230 (2005): 233–240; A. R. Koudinov and T. T. Berezov, "Alzheimer's Amyloid-Beta (Aβ) Is an Essential Synaptic Protein, Not Neurotoxic Junk," *Acta Neurobiologiae Experimentalis* 64 (2004): 71–79.

22. H. M. Lanoiselee et al., "APP, PSEN1, and PSEN2 Mutations in Early-Onset Alzheimer Disease: A Genetic Screening Study of Familial and Sporadic Cases," *PLOS Medicine* 14 (2017): e1002270.

23. A. A. Nugent et al., "TREM2 Regulates Microglial Cholesterol Metabolism upon Chronic Phagocytic Challenge," *Neuron* 105 (2020): 837–854.

24. G. Ponath et al., "Myelin Phagocytosis by Astrocytes after Myelin Damage Promotes Lesion Pathology," *Brain* 140 (2017): 399–413.

25. P. Padmanabham, S. Liu, and D. Silverman, "Lipophilic Statins in Subjects with Early Mild Cognitive Impairment: Associations with Conversion to Dementia and Decline in Posterior Cingulate Brain Metabolism in a Long-Term Prospective Longitudinal Multi-Center Study," *Journal of Nuclear Medicine* 62, suppl. 1 (2021): 102.

26. C. G. Fernandez et al., "The Role of APOE4 in Disrupting the Homeostatic Functions of Astrocytes and Microglia in Aging and Alzheimer's Disease," *Frontiers in Aging Neuroscience* 11 (2019): 14.

27. H. Jick et al., "Statins and the Risk of Dementia," *Lancet* 356 (2000): 1627–1631; B. Wolozin et al., "Decreased Prevalence of Alzheimer Disease Associated with 3-Hydroxy-3-Methyglutaryl Coenzyme A Reductase Inhibitors," *Archives of Neurology* 57 (2000): 1439–1443.

28. R. C. Petersen and S. Negash, "Mild Cognitive Impairment: An Overview," *CNS Spectrums* 13 (2008): 45–53.

29. S. C. McEwen et al., "A Systems-Biology Clinical Trial of a Personalized Multimodal Lifestyle Intervention for Early Alzheimer's Disease," *Alzheimer's & Dementia* 7 (2021): e12191.

9. Cancers at a Turning Point

1. C. Mattiuzzi and G. Lippi, "Current Cancer Epidemiology," *Journal of Epidemiology and Global Health* 9 (2019): 217–222.

2. S. J. Henley, E. M. Ward, S. Scott, J. Ma, R. N. Anderson, A. U. Firth, C. C. Thomas, F. Islami, H. K. Weir, D. R. Lewis, R. L. Sherman, M. Wu, V. B. Benard, L. C. Richardson, A. Jemal, K. Cronin, and B. A. Kohler, "Annual Report to the Nation on the Status of Cancer, Part I: National Cancer Statistics," *Cancer* 126 (2020): 2225–2249.

3. J. Horgan, "Sorry, but So Far War on Cancer Has Been a Bust," *Cross-Check* (blog), *Scientific American*, May 21, 2014, https://blogs.scientificamerican.com/cross-check/sorry-but-so-far-war-on-cancer-has-been-a-bust/.

4. Y. Sun et al., "Treatment-Induced Damage to the Tumor Microenvironment Promotes Prostate Cancer Therapy Resistance through WNT16B," *Nature Medicine* 18 (2012): 1359–1368; C. A. Schmitt, C. T. Rosenthal, and S. W.

Lowe, "Genetic Analysis of Chemoresistance in Primary Murine Lymphomas," *Nature Medicine* 6 (2000): 1029–1035; T. L. Wang et al., "Digital Karyotyping Identifies Thymidylate Synthase Amplification as a Mechanism of Resistance to 5-Fluorouracil in Metastatic Colorectal Cancer Patients," *Proceedings of the National Academy of Sciences of the United States of America* 101 (2004): 3089–3094; B. Campos et al., "A Comprehensive Profile of Recurrent Glioblastoma," *Oncogene* 35 (2016): 5819–5825.

5. Y. Song et al., "Evolutionary Etiology of High-Grade Astrocytomas," *Proceedings of the National Academy of Sciences of the United States of America* 110 (2013): 17933–17938.

6. C. Presutti, "HIV Drug Sped to Approval 25 Years Ago Revolutionized Fight against AIDS," *Voice of America,* June 22, 2020, https://www.voanews.com/a/usa_hiv-drug-sped-approval-25-years-ago-revolutionized-fight-against-aids/6191517.html.

7. M. A. Kutny et al., "Assessment of Arsenic Trioxide and All-Trans Retinoic Acid for the Treatment of Pediatric Acute Promyelocytic Leukemia: A Report from the Children's Oncology Group AAML1331 Trial," *JAMA Oncology* 8 (2022): 79–87.

8. B. Karai et al., "A Novel Flow Cytometric Method for Enhancing Acute Promyelocytic Leukemia Screening by Multidimensional Dot-Plots," *Annals of Hematology* 98 (2019): 1413–1420.

9. L. J. Old, "Cancer Immunology," *Scientific American* 236 (1977): 62–70, 72–73, 76, 79.

10. P. Dobosz and T. Dzieciatkowski, "The Intriguing History of Cancer Immunotherapy," *Frontiers in Immunology* 10 (2019): 2965.

11. L. Hood, "A Personal Journey of Discovery: Developing Technology and Changing Biology," *Annual Review of Analytical Chemistry* 1 (2008): 1–43.

12. D. R. Leach, M. F. Krummel, and J. P. Allison, "Enhancement of Antitumor Immunity by CTLA-4 Blockade," *Science* 271 (1996): 1734–1736.

13. Y. Ishida et al., "Induced Expression of PD-1, a Novel Member of the Immunoglobulin Gene Superfamily, upon Programmed Cell Death," *EMBO Journal* 11 (1992): 3887–3895.

14. D. Tseng et al., "Anti-CD47 Antibody-Mediated Phagocytosis of Cancer by Macrophages Primes an Effective Antitumor T-Cell Response," *Proceedings of the National Academy of Sciences of the United States of America* 110 (2013): 11103–11108.

15. S. A. Rosenberg et al., "Use of Tumor-Infiltrating Lymphocytes and Interleukin-2 in the Immunotherapy of Patients with Metastatic Melanoma: A Preliminary Report," *New England Journal of Medicine* 319 (1988): 1676–1680.

16. J. N. Brudno et al., "Safety and Feasibility of Anti-CD19 CAR T Cells with Fully Human Binding Domains in Patients with B-Cell Lymphoma," *Nature Medicine* 26 (2020): 270–280.

17. H. T. Marshall and M.B.A. Djamgoz, "Immuno-Oncology: Emerging Targets and Combination Therapies," *Frontiers in Oncology* 8 (2018): 315.

18. A. Hoos, "Development of Immuno-Oncology Drugs—From CTLA4 to PD1 to the Next Generations," *Nature Reviews Drug Discovery* 15 (2016): 235–247; W. C. M. Dempke et al., "Second- and Third-Generation Drugs for Immuno-Oncology Treatment—The More the Better?" *European Journal of Cancer* 74 (2017): 55–72.

19. S. Upadhaya, V. M. Hubbard-Lucey, and J. X. Yu, "Immuno-Oncology Drug Development Forges on Despite COVID-19," *Nature Reviews Drug Discovery* 19 (2020): 751–752.

20. J. Xin Yu, V. M. Hubbard-Lucey, and J. Tang, "Immuno-Oncology Drug Development Goes Global," *Nature Reviews Drug Discovery* 18 (2019): 899–900.

21. W. Tabayoyong and R. Abouassaly, "Prostate Cancer Screening and the Associated Controversy," *Surgical Clinics of North America* 95 (2015): 1023–1039.

22. M. C. Liu et al., "Sensitive and Specific Multi-Cancer Detection and Localization Using Methylation Signatures in Cell-Free DNA," *Annals of Oncology* 31 (2020): 745–759.

23. E. Klein et al., "Clinical Validation of a Targeted Methylation-Based Multi-Cancer Early Detection Test Using an Independent Validation Set." *Annals of Oncology* 32 (2021): 1167–1177.

24. A. T. Magis et al., "Untargeted Longitudinal Analysis of a Wellness Cohort Identifies Markers of Metastatic Cancer Years Prior to Diagnosis," *Scientific Reports* 10 (2020): 16275.

25. P. Kearney et al., "The Building Blocks of Successful Translation of Proteomics to the Clinic," *Current Opinion in Biotechnology* 51 (2018): 123–129.

26. X. J. Li et al., "A Blood-Based Proteomic Classifier for the Molecular Characterization of Pulmonary Nodules," *Science Translational Medicine* 5 (2013): 207ra142.

27. G. A. Silvestri et al., "Assessment of Plasma Proteomics Biomarker's Ability to Distinguish Benign from Malignant Lung Nodules: Results of the Panoptic (Pulmonary Nodule Plasma Proteomic Classifier) Trial," *CHEST* 154 (2018): 491–500.

28. Li et al., "A Blood-Based Proteomic Classifier."

29. Kearney et al., "The Building Blocks of Successful Translation."

30. Disclosure: We have both served as scientific advisors to Sera Prognostics for many years and hold stock options in the company. Sera went public on the NASDAQ in 2021.

31. D. F. Hayes, "HER2 and Breast Cancer—A Phenomenal Success Story," *New England Journal of Medicine* 381 (2019): 1284–1286.

32. M. Marty et al., "Randomized Phase II Trial of the Efficacy and Safety of Trastuzumab Combined with Docetaxel in Patients with Human Epidermal Growth Factor Receptor 2-Positive Metastatic Breast Cancer Administered as First-Line Treatment: The M77001 Study Group," *Journal of Clinical Oncology* 23 (2005): 4265–4274; M. J. Piccart-Gebhart et al., "Trastuzumab after Adjuvant Chemotherapy in HER2-Positive Breast Cancer," *New England Journal of Medicine* 353 (2005): 1659–1672.

33. J. T. Jorgensen et al., "A Companion Diagnostic with Significant Clinical Impact in Treatment of Breast and Gastric Cancer," *Frontiers in Oncology* 11 (2021): 676939.

34. V. Valla et al., "Companion Diagnostics: State of the Art and New Regulations," *Biomarker Insights* 16 (2021): 11772719211047763.

35. M. Gromova et al., "Biomarkers: Opportunities and Challenges for Drug Development in the Current Regulatory Landscape," *Biomarker Insights* 15 (2020): 1177271920974652.

36. S. Shen, S. Vagner, and C. Robert, "Persistent Cancer Cells: The Deadly Survivors," *Cell* 183 (2020): 860–874.

37. American Cancer Society, "Cancer Statistics Center," 2021, https://cancerstatisticscenter.cancer.org/#!/.

38. J. Marquart, E. Y. Chen, and V. Prasad, "Estimation of the Percentage of US Patients with Cancer Who Benefit from Genome-Driven Oncology," *JAMA Oncology* 4 (2018): 1093–1098; A. Haslam, M. S. Kim, and V. Prasad, "Updated Estimates of Eligibility for and Response to Genome-Targeted Oncology Drugs among US Cancer Patients, 2006–2020," *Annals of Oncology* 32 (2021): 926–932.

39. D. L. Jardim et al., "The Challenges of Tumor Mutational Burden as an Immunotherapy Biomarker," *Cancer Cell* 39 (2021): 154–173.

40. Members of the Oncology Think Tank, "We Must Find Ways to Detect Cancer Much Earlier," *Scientific American*, January 8, 2021, https://www.scientificamerican.com/article/we-must-find-ways-to-detect-cancer-much-earlier/.

41. Nathan served on the Board of Advisors for the Pacific Northwest chapter of the American Cancer Society.

10. The AI Imperative

1. J. Pearl and D. Mackenzie, *The Book of Why* (Penguin Books, 2019).
2. G. Kasparov, "The Chess Master and the Computer," *New York Review*, February 11, 2010.
3. R. F. Service, "'The Game Has Changed.' AI Triumphs at Protein Folding," *Science* 370 (2020): 1144–1145.
4. M. Baek et al., "Accurate Prediction of Protein Structures and Interactions Using a Three-Track Neural Network," *Science* 373 (2021): 871–876.
5. J. C. Phillips et al., "Scalable Molecular Dynamics on CPU and GPU Architectures with NAMD," *Journal of Chemical Physics* 153 (2020): 044130; X. Liu et al., "Molecular Dynamics Simulations and Novel Drug Discovery," *Expert Opinion on Drug Discovery* 13 (2018): 23–37; Y. Wang, J. M. Lamim Ribeiro, and P. Tiwary, "Machine Learning Approaches for Analyzing and Enhancing Molecular Dynamics Simulations," *Current Opinion in Structural Biology* 61 (2020): 139–145.
6. D. O. Hebb, *The Organization of Behavior: A Neuropsychological Theory* (Wiley and Sons, 1949).
7. M. D. Huesch and T. J. Mosher, "Using It or Losing It? The Case for Data Scientists inside Health Care," *NEJM Catalyst*, May 4, 2017, https://catalyst.nejm.org/case-data-scientists-inside-health-care/.
8. S. Nuti and M. Vainieri, "Managing Waiting Times in Diagnostic Medical Imaging," *BMJ Open* 2 (2012).
9. R. Yousef et al., "A Holistic Overview of Deep Learning Approach in Medical Imaging," *Multimedia Systems* 28 (2022): 881–914; S. K. Zhou et al., "Deep Reinforcement Learning in Medical Imaging: A Literature Review," *Medical Image Analysis* 73 (2021): 102193; A. Esteva et al., "A Guide to Deep Learning in Healthcare," *Nature Medicine* 25 (2019): 24–29.
10. M. S. Kim et al., "Artificial Intelligence and Lung Cancer Treatment Decision: Agreement with Recommendation of Multidisciplinary Tumor Board," *Translational Lung Cancer Research* 9 (2020): 507–514; A. Mitani et al., "Detection of Anaemia from Retinal Fundus Images via Deep Learning," *Nature Biomedical Engineering* 4 (2020): 18–27; S. M. McKinney et al., "International Evaluation of an AI System for Breast Cancer Screening," *Nature* 577 (2020): 89–94; O. J. Oktay et al., "Evaluation of Deep Learning to Augment Image-Guided Radiotherapy for Head and Neck and Prostate Cancers," *JAMA Network Open* 3 (2020): e2027426.
11. V. Gulshan et al., "Development and Validation of a Deep Learning Algorithm for Detection of Diabetic Retinopathy in Retinal Fundus Photographs," *Journal of the American Medical Association* 316 (2016): 2402–2410.

12. A. Y. Hannun et al., "Cardiologist-Level Arrhythmia Detection and Classification in Ambulatory Electrocardiograms Using a Deep Neural Network," *Nature Medicine* 25 (2019): 65–69.

13. L. Moja et al., "Effectiveness of Computerized Decision Support Systems Linked to Electronic Health Records: A Systematic Review and Meta-Analysis," *American Journal of Public Health* 104 (2014): e12–e22.

14. J. G. Anderson and K. Abrahamson, "Your Health Care May Kill You: Medical Errors," *Studies in Health Technology and Informatics* 234 (2017): 13–17.

15. Anderson and Abrahamson, "Your Health Care May Kill You."

16. E. J. Topol, "High-Performance Medicine: The Convergence of Human and Artificial Intelligence," *Nature Medicine* 25 (2019): 44–56; A. Haque, A. Milstein, and L. Fei-Fei, "Illuminating the Dark Spaces of Healthcare with Ambient Intelligence," *Nature* 585 (2020): 193–202; R. T. Sutton et al., "An Overview of Clinical Decision Support Systems: Benefits, Risks, and Strategies for Success," *NPJ Digital Medicine* 3 (2020): 17.

17. R. Rozenblum et al., "Using a Machine Learning System to Identify and Prevent Medication Prescribing Errors: A Clinical and Cost Analysis Evaluation," *Joint Commission Journal on Quality and Patient Safety* 46 (2020): 3–10.

18. G. P. Velo and P. Minuz, "Medication Errors: Prescribing Faults and Prescription Errors," *British Journal of Clinical Pharmacology* 67 (2009): 624–628.

19. Rozenblum et al., "Using a Machine Learning System."

20. E. R. Doherty-Torstrick, K. E. Walton, and B. A. Fallon, "Cyberchondria: Parsing Health Anxiety from Online Behavior," *Psychosomatics* 57 (2016): 390–400.

21. C. Metz, "AI Is Transforming Google Search: The Rest of the Web Is Next," *Wired*, February 4, 2016, https://www.wired.com/2016/02/ai-is-changing-the-technology-behind-google-searches/.

22. Sutton et al., "An Overview of Clinical Decision Support Systems."

23. S. M. Kabene, *Healthcare and the Effect of Technology: Developments, Challenges and Advancements* (Medical Information Science Reference, 2010).

24. Sutton et al., "An Overview of Clinical Decision Support Systems."

25. P. J. Embi et al., "Development of an Electronic Health Record-Based Clinical Trial Alert System to Enhance Recruitment at the Point of Care," AMIA *Annual Symposium Proceedings* (2005): 231–235.

26. D. McEvoy et al., "Enhancing Problem List Documentation in Electronic Health Records Using Two Methods: The Example of Prior Splenectomy," *BMJ Quality & Safety* 27 (2018): 40–47.

27. R. Kunhimangalam, S. Ovallath, and P. K. Joseph, "A Clinical Decision Support System with an Integrated EMR for Diagnosis of Peripheral Neuropathy," *Journal of Medical Systems* 38 (2014): 38.

28. A. I. Martinez-Franco et al., "Diagnostic Accuracy in Family Medicine Residents Using a Clinical Decision Support System (DXplain): A Randomized-Controlled Trial," *Diagnosis* 5 (2018): 71–76.

29. B. Keltch, Y. Lin, and C. Bayrak, "Comparison of AI Techniques for Prediction of Liver Fibrosis in Hepatitis Patients," *Journal of Medical Systems* 38 (2014): 60.

30. L. Morkrid et al., "Continuous Age- and Sex-Adjusted Reference Intervals of Urinary Markers for Cerebral Creatine Deficiency Syndromes: A Novel Approach to the Definition of Reference Intervals," *Clinical Chemistry* 61 (2015): 760–768.

31. P. Spyridonos et al., "A Computer-Based Diagnostic and Prognostic System for Assessing Urinary Bladder Tumour Grade and Predicting Cancer Recurrence," *Medical Informatics and the Internet in Medicine* 27 (2002): 111–122; E. Tsolaki et al., "Fast Spectroscopic Multiple Analysis (FASMA) for Brain Tumor Classification: A Clinical Decision Support System Utilizing Multi-Parametric 3T MR Data," *International Journal of Computer Assisted Radiology and Surgery* 10 (2015): 1149–1166.

32. G. Coorey et al., "The Health Digital Twin: Advancing Precision Cardiovascular Medicine," *Nature Reviews Cardiology* 18 (2021): 803–804; T. Hernandez-Boussard et al., "Digital Twins for Predictive Oncology Will Be a Paradigm Shift for Precision Cancer Care," *Nature Medicine* 27 (2021): 2065–2066.

33. Coorey et al., "The Health Digital Twin"; Hernandez-Boussard et al., "Digital Twins for Predictive Oncology."

34. M. N. Kamel Boulos and P. Zhang, "Digital Twins: From Personalised Medicine to Precision Public Health," *Journal of Personalized Medicine* 11 (2021).

35. P. H. Huang, K. H. Kim, and M. Schermer, "Ethical Issues of Digital Twins for Personalized Health Care Service: Preliminary Mapping Study," *Journal of Medical Internet Research* 24 (2022): e33081; E. O. Popa et al., "The Use of Digital Twins in Healthcare: Socio-Ethical Benefits and Socio-Ethical Risks," *Life Sciences, Society and Policy* 17 (2021): 6.

11. The Path Forward

1. Centers for Disease Control and Prevention, National Center for Health Statistics, "Percent of U.S. Adults 55 and Over with Chronic Conditions," last reviewed November 6, 2015, https://www.cdc.gov/nchs/health_policy/adult_chronic_conditions.htm.

2. Y. Su et al., "Multiple Early Factors Anticipate Post-Acute COVID-19 Sequelae," *Cell* 185 (2022): 881–895.e20; Y. Su et al., "Multi-Omics Resolves

a Sharp Disease-State Shift between Mild and Moderate COVID-19," *Cell* 183 (2020): 1479–1495.e20; J. W. Lee et al., "Integrated Analysis of Plasma and Single Immune Cells Uncovers Metabolic Changes in Individuals with COVID-19," *Nature Biotechnology* 40 (2022): 110–120.

3. Su et al., "Multiple Early Factors"; Su et al., "Multi-Omics Resolves a Sharp Disease-State Shift"; Lee et al., "Integrated Analysis of Plasma and Single Immune Cells."

4. M. Treskova-Schwarzbach et al., "Pre-Existing Health Conditions and Severe COVID-19 Outcomes: An Umbrella Review Approach and Meta-Analysis of Global Evidence," *BMC Medicine* 19 (2021): 212.

5. J. J. Zhang et al., "Risk and Protective Factors for COVID-19 Morbidity, Severity, and Mortality," *Clinical Reviews in Allergy & Immunology* (2022): 1–18.

6. W. B. Grant, "Vitamin D's Role in Reducing Risk of SARS-CoV-2 and COVID-19 Incidence, Severity, and Death," *Nutrients* 14 (2022): 183.

7. H. Shakoor et al., "Immune-Boosting Role of Vitamins D, C, E, Zinc, Selenium and Omega-3 Fatty Acids: Could They Help against COVID-19?" *Maturitas* 143 (2021): 1–9.

8. H. Onal et al., "Treatment of COVID-19 Patients with Quercetin: A Prospective, Single Center, Randomized, Controlled Trial," *Turkish Journal of Biology* 45 (2021): 518–529; F. Di Pierro et al., "Potential Clinical Benefits of Quercetin in the Early Stage of COVID-19: Results of a Second, Pilot, Randomized, Controlled and Open-Label Clinical Trial," *International Journal of General Medicine* 14 (2021): 2807–2816; F. Di Pierro et al., "Possible Therapeutic Effects of Adjuvant Quercetin Supplementation against Early-Stage COVID-19 Infection: A Prospective, Randomized, Controlled, and Open-Label Study," *International Journal of General Medicine* 14 (2021): 2359–2366; A. Saeedi-Boroujeni and M. R. Mahmoudian-Sani, "Anti-Inflammatory Potential of Quercetin in COVID-19 Treatment," *Journal of Inflammation* 18 (2021): 3; F. Di Pierro et al., "Quercetin Phytosome® as a Potential Candidate for Managing COVID-19," *Minerva Gastroenterology* 67 (2021): 190–195; S. Bastaminejad and S. Bakhtiyari, "Quercetin and Its Relative Therapeutic Potential against COVID-19: A Retrospective Review and Prospective Overview," *Current Molecular Medicine* 21 (2021): 385–391; M. Aucoin et al., "The Effect of Quercetin on the Prevention or Treatment of COVID-19 and Other Respiratory Tract Infections in Humans: A Rapid Review," *Advances in Integrative Medicine* 7 (2020): 247–251.

9. G. D. Batty et al., "Explaining Ethnic Differentials in COVID-19 Mortality: A Cohort Study," *American Journal of Epidemiology* 191 (2022): 275–281.

10. S. Magesh et al., "Disparities in COVID-19 Outcomes by Race, Ethnicity, and Socioeconomic Status: A Systematic-Review and Meta-Analysis," *JAMA Network Open* 4 (2021): e2134147.

11. C. A. Taylor et al., "COVID-19-Associated Hospitalizations among Adults During SARS-CoV-2 Delta and Omicron Variant Predominance, by Race / Ethnicity and Vaccination Status—COVID-NET, 14 States, July 2021–January 2022," *Morbidity and Mortality Weekly Report* 71 (2022): 466–473.

12. D. E. Willis, "COVID-19 Vaccine Hesitancy: Race / Ethnicity, Trust, and Fear," *Clinical and Translational Science* 14 (2021): 2200–2207.

13. A. Raharja, A. Tamara, and L. T. Kok, "Association between Ethnicity and Severe COVID-19 Disease: A Systematic Review and Meta-Analysis," *Journal of Racial and Ethnic Health Disparities* 8 (2021): 1563–1572.

14. S. Richards et al., "Standards and Guidelines for the Interpretation of Sequence Variants: A Joint Consensus Recommendation of the American College of Medical Genetics and Genomics and the Association for Molecular Pathology," *Genetics in Medicine* 17 (2015): 405–424; D. T. Miller et al., "ACMG SF v3.0 List for Reporting of Secondary Findings in Clinical Exome and Genome Sequencing: A Policy Statement of the American College of Medical Genetics and Genomics (ACMG)," *Genetics in Medicine* 23 (2021): 1381–1390.

15. C. Hippman and C. Nislow, "Pharmacogenomic Testing: Clinical Evidence and Implementation Challenges," *Journal of Personalized Medicine* 9 (2019).

16. M. Wainberg et al., "Multiomic Blood Correlates of Genetic Risk Identify Presymptomatic Disease Alterations," *Proceedings of the National Academy of Sciences of the United States of America* 117 (2020): 21813–21820.

17. Polygenic Risk Score Task Force of the International Common Disease Alliance, "Responsible Use of Polygenic Risk Scores in the Clinic: Potential Benefits, Risks and Gaps," *Nature Medicine* 27 (2021): 1876–1884; A. C. F. Lewis and R.C. Green, "Polygenic Risk Scores in the Clinic: New Perspectives Needed on Familiar Ethical Issues," *Genome Medicine* 13 (2021): 14.

18. J. C. Roach et al., "Analysis of Genetic Inheritance in a Family Quartet by Whole-Genome Sequencing," *Science* 328 (2010): 636–639.

19. R. Sakate and T. Kimura, "Drug Repositioning Trends in Rare and Intractable Diseases," *Drug Discovery Today* 27 (2022): 1789–1795.

20. H. I. Roessler et al., "Drug Repurposing for Rare Diseases," *Trends in Pharmacological Sciences* 42 (2021): 255–267.

21. M. Heron, "Deaths: Leading Causes for 2016," *National Vital Statistics Reports* 67 (2018): 1–77.

22. J. S. Roberts et al., "Direct-to-Consumer Genetic Testing: User Motivations, Decision Making, and Perceived Utility of Results," *Public Health Genomics* 20 (2017): 36–45.

23. K. D. Christensen et al., "A Randomized Controlled Trial of Disclosing Genetic Risk Information for Alzheimer Disease via Telephone," *Genetics in Medicine* 20 (2018): 132–141.

24. L. M. Amendola, K. Golden-Grant, and S. Scollon, "Scaling Genetic Counseling in the Genomics Era," *Annual Review of Genomics and Human Genetics* 22 (2021): 339–355.

25. I. A. Holm et al., "The BabySeq Project: Implementing Genomic Sequencing in Newborns," *BMC Pediatrics* 18 (2018): 225.

26. B. Armstrong et al., "Parental Attitudes toward Standard Newborn Screening and Newborn Genomic Sequencing: Findings from the BabySeq Study," *Frontiers in Genetics* 13 (2022): 867371; M. H. Wojcik et al., "Discordant Results between Conventional Newborn Screening and Genomic Sequencing in the BabySeq Project," *Genetics in Medicine* 23 (2021): 1372–1375; S. Pereira et al., "Perceived Benefits, Risks, and Utility of Newborn Genomic Sequencing in the BabySeq Project," *Pediatrics* 143 (2019): S6–S13; I. A. Holm et al., "Returning a Genomic Result for an Adult-Onset Condition to the Parents of a Newborn: Insights from the BabySeq Project," *Pediatrics* 143 (2019): S37–S43; C. A. Genetti et al., "Parental Interest in Genomic Sequencing of Newborns: Enrollment Experience from the BabySeq Project," *Genetics in Medicine* 21 (2019): 622–630; O. Ceyhan-Birsoy et al., "Interpretation of Genomic Sequencing Results in Healthy and Ill Newborns: Results from the BabySeq Project," *American Journal of Human Genetics* 104 (2019): 76–93.

27. D. Crosby, "Delivering on the Promise of Early Detection with Liquid Biopsies," *British Journal of Cancer* 126 (2022): 313–315; A. Markou et al., "DNA Methylation Analysis of Tumor Suppressor Genes in Liquid Biopsy Components of Early Stage NSCLC: A Promising Tool for Early Detection," *Clinical Epigenetics* 14 (2022): 61; M. Shoukry et al., "The Emerging Role of Circulating Tumor DNA in the Management of Breast Cancer," *Cancers* 13 (2021); M. Nagasaka et al., "Liquid Biopsy for Therapy Monitoring in Early-Stage Non-Small Cell Lung Cancer," *Molecular Cancer* 20 (2021): 82; C. Rolfo and A. Russo, "Liquid Biopsy for Early Stage Lung Cancer Moves Ever Closer," *Nature Reviews Clinical Oncology* 17 (2020): 523–524.

28. D. S. Haslem et al., "A Retrospective Analysis of Precision Medicine Outcomes in Patients with Advanced Cancer Reveals Improved Progression-Free Survival without Increased Health Care Costs," *Journal of Oncology Practice* 13 (2017): e108–e119; D. S. Haslem et al., "Precision Oncology in Advanced Cancer Patients Improves Overall Survival with Lower Weekly Healthcare Costs," *Oncotarget* 9 (2018): 12316–12322.

29. P. M. Matthews and C. Sudlow, "The UK Biobank," *Brain* 138 (2015): 3463–3465; C. Sudlow et al., "UK Biobank: An Open Access Resource for Identifying the Causes of a Wide Range of Complex Diseases of Middle and Old Age," *PLOS Medicine* 12 (2015): e1001779.

30. B. B. Sun et al., "Genetic Regulation of the Human Plasma Proteome in 54,306 UK Biobank Participants," *bioRxiv* (2022): 2022.2006.2017.496443.

31. The All of Us Research Program Investigators, "The 'All of Us' Research Program," *New England Journal of Medicine* 381 (2019): 668–676.

32. J. M. Chapel et al., "Prevalence and Medical Costs of Chronic Diseases among Adult Medicaid Beneficiaries," *American Journal of Preventive Medicine* 53 (2017): S143–S154; N. J. Schork, "Personalized Medicine: Time for One-Person Trials," *Nature* 520 (2015): 609–611.

33. K. Gebreyes et al., "Breaking the Cost Curve," Deloitte Insights, February 9, 2021, https://www2.deloitte.com/xe/en/insights/industry/health-care/future-health-care-spending.html.

34. W. Ji et al., "Wearable Sweat Biosensors Refresh Personalized Health/Medical Diagnostics," *Research* (2021): 9757126.

35. A. Borsky et al., "Few Americans Receive All High-Priority, Appropriate Clinical Preventive Services," *Health Affairs* 37 (2018): 925–928.

36. R. A. Beckman and C. Chen, "New Evidence-Based Adaptive Clinical Trial Methods for Optimally Integrating Predictive Biomarkers into Oncology Clinical Development Programs," *Chinese Journal of Cancer* 32 (2013): 233–241.

37. J. T. Yurkovich et al., "A Systems Approach to Clinical Oncology Uses Deep Phenotyping to Deliver Personalized Care," *Nature Reviews Clinical Oncology* 17 (2020): 183–194.

Acknowledgments

We would like to thank the many people who have helped us on our way to scientific wellness and the writing of this book. Our thanks are by no means complete. It truly takes a village to formulate a new vision of healthcare and push it toward reality.

First of all, we'd like to thank Matthew LaPlante, whose editing and insightful modifications elevated and clarified the prose and were instrumental to making this book accessible. We appreciate his patience and willingness to go the extra mile throughout this process. He was a wonderful partner in the development of this book, and his role was fundamental to its readability.

Our heartfelt gratitude, also, to our Harvard University Press editor, Joy de Menil, who made many important conceptual and editorial contributions. We are very grateful for the time and care she put into this effort, for her insistence that more of the science be included (but hopefully not too much!), and for her help in clarifying often complex concepts. She really improved our book.

A special thanks to Thom Mount, who inspired us to start this project and who has been a tireless help to us in crafting more potent messaging throughout the several years of writing. With his background as former president of Universal Pictures and a lifetime in the movie business, he challenged us to work clearer narratives into sharing a message to make it accessible to all.

Thank you also to Becky Bogard for her skepticism and clarification, Jeff Zimman, with his fantastic journalistic writing chops, for help with Chapter 6, and Simon Evans for additions related to nutrition and the brain. Thank you to Gil Omenn for critical early feedback and key advice throughout our whole journey, Tom Patterson for conceptual input and the illustrations for Chapter 8, Jared Roach for details on Alzheimer's clinical trials, John Earls and Andrew Magis for their insights into health AI in Chapter 9, Allison Kudla for her illustration artistry, Paul Lange for his thoughtful contributions, and Susan Paynter for her keen eye in making line edits and thoughtful comments throughout.

Many people helped in crucial ways with the research and work upon which the book is based. We'd like to thank our Arivale colleagues and the members who allowed their de-identified data to be used for scientific discovery. The quality and commitment of this team to the future of health—from the wonderful Arivale coaches to the science and development team and all the business functions in between—were inspiring. We feel privileged to have been associated with each one of you. Of particular note from the earliest days are Sean Bell, Kristin Brogaard, John Earls, Sandi Kaplan, Clayton Lewis, Jennifer Lovejoy, Andrew Magis, Sara Mecca, and Mia Neese, as well as all the participants in the foundational Pioneer 100 wellness study. Special thanks also to our personal health coaches while we were going through the Arivale program ourselves, Jessica Roberts and Ginger Hultin. We owe you all an immense debt of gratitude. Health data discussed in this book are in many cases attributed to individuals by a first name only. In those cases, the names have been changed.

We'd also like to thank all of the members of the Hood-Price Lab who worked with us to pioneer the science described in the book and in so many other areas as well. It has been a pleasure and an honor to work with such a high-caliber group of people from whom we learn so much every day. Special thanks to those who did the data analysis work, including huge contributions across many years from Noa Rappaport, Tomasz Wilmanski, Andrew Magis, and John Earls. Special thanks also to our ISB colleagues who have helped to push forward this science, especially Sean Gibbons, Jenn Hadlock, Jim Heath, Sui Huang, and Ilya Shmulevich. Also, special thanks to Kalli Trachana, who was such an important part of working to translate these ideas to the world.

Lee is grateful to the early members of newly created nonprofit Phenome Health for helping to build the team for a million-person Human Phenome Initiative, to articulate the vision, begin to implement the strategy, develop the education programs, recruit critical partners, and lobby Congress. These include Jeff Boore, Simon Evans, Lisa Kamemoto,

Robbie Kilpatrick, Charles Richardson, Tim Yeatman, BJ Yurkovich, and James Yurkovich, who are involved in the science, computation, communications, and administration. Becky Bogard, Rick Desimone, Liz Fortunato, and Tim Zenk are our invaluable local and federal lobbyists. We also thank the twelve members of the Phenome Health Senior Advisory Board for their insightful counsel and guidance.

Nathan is grateful to the many terrific colleagues who are working together to translate personalized scientific wellness into practice at Thorne HealthTech. First and foremost, Paul Jacobson is a visionary leader with a rare talent for combining a no-nonsense practical business sense with the boldness to venture into new areas. A special thanks to Bodi Zhang, a partner in all of this, for recognizing key strategies and detailed implementation needed to take an idea and turn it into a reality. Huge appreciation for Tom McKenna, whose tireless work ethic and strong leadership have earned him universal respect and who makes so much of the company go, and Will McCamy, whose energy and drive make such a difference in getting solutions out to the world. It has been an immense pleasure and honor to work with so many exceptional people on the Thorne science and medical teams, including Joel Dudley, John Earls, Amanda Frick, Jacqueline Jacques, Loukia Lili, Chris Mason, Cem Meydan, Sarah Pesce, Stephen Phipps, Ben Readhead, Mary Kay Ross, Bob Rountree, Jerome Scelza, Caleb Schmidt, Michael Schmidt, and many others. Thank you also to Tamarah Strauss and all the team members who help to promote personalized scientific wellness to the world.

Thank you also to many people who have influenced our thinking along our journey, including Jeff Bland, George Church, Joel Dudley, Sara Gottfried, George Haddad, Rod Hochman, Michele Leary, Mike Merzenich, Craig Mundie, Tom Paterson, Scott Penberthy, Roger Perlmutter, Jen Rohrs, Dave Sabey, Suchi Saria, Michael Schwartz, Ralph Snyderman, Chuck Watts, and Jeff Wilke.

Thanks to all the members of the Logan Center for Education at ISB, especially Caroline Kiehle, its director, who has been working with Lee for the past thirty years guiding our adventures in transforming K–12 science education—a critical component for all modern science.

Finally, we want to thank to our families.

Lee: I would like to give a very special thanks to my wife, Valerie Logan, who started us down the path of K–12 science education leading to the creation of the Logan Center for Education. My children, Eran and Marqui, were a constant touchstone to the reality of non-science readers throughout the emergence of this book. No father could be more proud of the wonderful adults his children have become. I also want to thank Joanne Fiorito for the loving care she has given Valerie over these past ten years—she is really a part of our family. Becky Bogard, my recent partner, has been incredibly contributory and accommodating for this hectic and complex process of writing and repeatedly rewriting our book.

Nathan: I am especially grateful to my wife, Brenda, for all her love and support and for enabling me to spend so much time over evenings, weekends, and the long retreats required to finish the book. And to my children, Camille, Madeline, and Sophia, for being a joy in my life. As you continue to grow up, I hope the age of scientific wellness will make the world a better place for you.

Index

Page numbers in *italics* refer to illustrations.